PLANT LORE
OF AN
ALASKAN ISLAND

PLANT LORE
OF AN
ALASKAN ISLAND

FRANCES KELSO GRAHAM
AND THE OUZINKIE BOTANICAL SOCIETY

ALASKA NORTHWEST PUBLISHING COMPANY
Anchorage, Alaska

Copyright © 1985 by Ouzinkie Botanical Society. All rights reserved. No part of this book may be reproduced or transmitted in any form or by any means, electronic or mechanical, including photocopying, recording or by any information storage and retrieval system, without written permission of Alaska Northwest Publishing Company, Box 4 EEE, Anchorage, Alaska 99509.

Library of Congress cataloging-in-publication data:
Graham, Frances Kelso, 1938-
   Plant lore of an Alaskan island.
    Bibliography: p. 184.
    Includes index.
    1. Wild plants, Edible — Alaska — Spruce Island.
2. Medicinal plants — Alaska — Spruce Island. 3. Cookery (Wild foods) 4. Dye plants — Alaska — Spruce Island.
5. Botany — Alaska — Spruce Island — Folklore.
6. Herbs — Alaska — Spruce Island. I. Ouzinkie Botanical Society. II. Title.
QK98.5.U6G73   1985   581.6′1′09798   85-15637
ISBN 0-88240-303-6

Illustrations by Sandra Coen
Illustration facing the title page by Janet Quaccia
Design by Sandra Harner
Cartography by Jon.Hersh

Alaska Northwest Publishing Company
Printed in U.S.A.

*To all the
Ouzinkie babushkas*

OUZINKIE BOTANICAL SOCIETY
EDITORIAL STAFF

Fran Kelso: Chief Flunky and Word Arranger

Sandra Coen: Illustrator

Georgia Smith: Typist Extraordinaire, Chief Assistant and Recipe Editor

Stacy Studebaker: Botanical Editor

Angeline Anderson: Oral Researcher, Fine Forager, and Great Cook

Eileen Boskofsky: Welcome Relief Typist and Word Processor Wizard

Jenny Chernikoff: Local Use Consultant

Alexandra (Sasha) Smith: Local Use Consultant, Field Trip Enthusiast and Class Mascot

Rosemary Squartsoff: Recipe Expert and Gatherer (knows the woods so well we call her "Bog Rosemary")

Greg Wolfer: Research Professor

Chris Quick, Sheila Anderson, Carl M. Smith, Thelma Anderson, Nell Tsacrios, Claudia Torsen: Past-Year Class Members (They each added something!)

Marian Brown, Nina Gilbreath, Father Gerasim, N.V.A: Russian Language Consultants

Janet Quaccia: Artist

Janey Kenyon: Preliminary Graphics

Daniel Konigsberg: Photographer

Ted Panamarioff: Preliminary Paste-Up Assistant

Roger Page: Preliminary Lay-Out and Printing Advisor (The editor's editor)

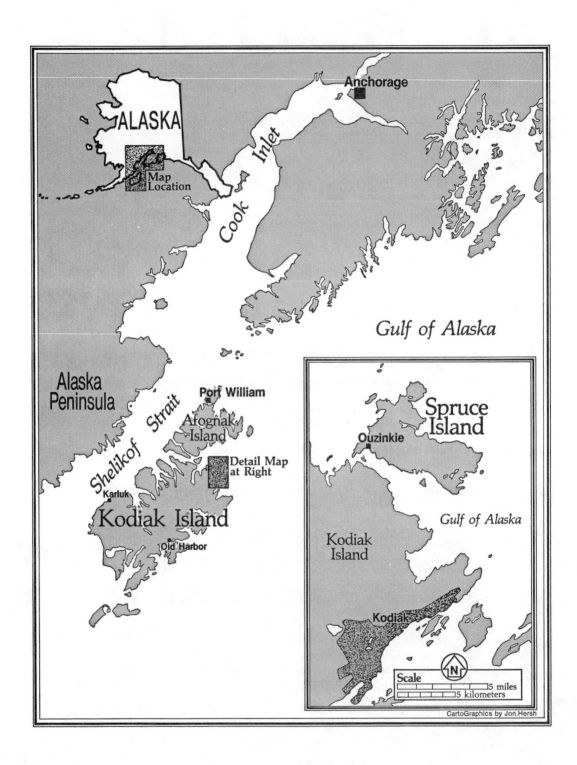

TABLE OF CONTENTS

Preface................................ix
Introduction: Using this Book............1

THE PLANTS
A Gathering of Herbs...................9
Ferns.................................82
Berries, Low and High.................87
Shrubs...............................118
Trees................................128
Poisonous Plants.....................134

ADDITIONAL COOKERY
Hints on Cooking Wild Edibles........143
Wild Game Recipes....................147
Seafood Recipes......................151

SPRUCE ISLAND TEA RECIPES.......162

DYES FROM WILD PLANTS..........164

A CONVERSATION WITH
SASHA AND JENNY................172

Glossary.............................181
Bibliography.........................184
Index................................187

PREFACE

This little book is the product of three years of study. In the autumn of 1980, I was working as an adult educator in the village of Ouzinkie (pronounced "you-zinc-ee"), Alaska, located near Kodiak on Spruce Island. As part of my job, I started a class on wild and edible plants. I knew a little about the subject and had long wanted to learn more, especially about the wild plants in our part of Alaska. A few other women in the village had also expressed an interest in the subject, so I went to my employers on the Adult Education staff at Kodiak Community College and asked them if they would be willing to sponsor the class.

Under the far-seeing and very capable direction of Carol Hagel at Kodiak Community College, the Adult Education Department had started what I felt was a very strong educational program for adults in the villages of the Kodiak area. I was embarking on my second year as an A.B.E. (Adult Basic Education) teacher in Ouzinkie, and had discovered my job to be challenging and stimulating. The coordinator of our Kodiak area village programs, Lynda Ritacco-Maciak, and my "young whippersnapper boss," John Mahoney, who traveled the village circuit to help us keep our programs running smoothly, and who approved new classes and workshops, were marvelous. They encouraged A.B.E. teachers to try new classes that village adults wanted to take. They gave me the OK to start "Plants Class" — the Ouzinkie Botanical Society.

There were nine of us, off and on, that first year. Each of us knew a little; some more than others. We started by sharing our knowledge and learning from each other. Then there were hours together poring over the Cooperative Extension Service's *Wild, Edible, and Poisonous Plants*, and *The Alaska-Yukon Wild Flowers Guide*, from Alaska Northwest Publishing Company. We'd see a plant that one or more of us would recognize; then we'd read about it and discuss it. With the help of Kodiak Community College and through our own resources, we began to collect other books which contained information about the plants we were discovering. We read these and reported our findings to class.

We began making field trips as soon as new plants came up in the spring. Experimenting with cooking wild edibles led to monthly potlucks where we utilized wild fish, game, and plant life in our dishes as much as possible. We began drinking teas, made from plants foraged in the woods, to relieve simple ailments. We grew in our knowledge of the abundant natural setting in which we lived and we gained a comradeship with each other through a shared love of nature. And finally, we gathered together what we learned and wrote this book, in the hope that other adults might wish to try a similar learning experience. Perhaps other Alaskans will add our guide to those they use to learn about the countryside in which they live.

One of the nice things the college did for our class was to pay for a visit to Ouzinkie by a guest lecturer from time to time. John Mahoney told me about Stacy Studebaker, a Kodiak teacher who taught a class similar to ours at Kodiak Community College, and who had started a herbarium, or plant library, there. I asked if she could come over to give a lecture for our class, and in February of that year she arrived for her first visit. A bad storm kept her on Spruce Island with us for four days, the wind chill factor hit minus 40 degrees, and she spent a day and a half helping me cut and haul enough wood to keep the cabin warm.

Stacy has helped our class considerably with plant identification, and has shared much useful information about plants with us. She and her Kodiak Community College class members have submitted some of their recipes for wild edibles for inclusion in the book. We are very happy about the contributions Stacy and her students have made.

We are especially pleased to introduce one of Stacy's students as the illustrator of our Alaskan herbal. When Stacy suggested asking Sandra Coen to do the drawings, and sent me an example of her work, I could hardly wait to call Sandra and talk to her about the project. Sandra's art background and her outstanding ability in plant identification, combined with a wonderful drawing talent, make these illustrations a charming and accurate contribution to this book. These drawings will be a tremendous help to the beginner who wishes to learn to identify Alaskan plants.

Thanks are due to many people whose energy and expertise, and fine publications, helped make this book possible. A few special "thanks" are mentioned here. (See the back of this book for a complete bibliography.) First, we are indebted to Eric Hulten, whose *Flora of Alaska and Neighboring Territories* was our "plant bible." And, we are especially grateful for an informative book titled *Dena'ina K'et'una, Tanaina Plantlore,* compiled by Priscilla Russell Kari and published by the Adult Literacy Laboratory of the University of Alaska. Ms. Kari's ethnobotany of the Tanaina people has been an extremely valuable source of data on the uses of wild plants by other Alaskans. It has also served as a comparison study as we have learned the ways in which people in our own area have used these plants. Glen Ray's *Root, Stem and Leaf: Wild Vegetables of Southeast Alaska* was another very useful Alaskan study.

The fine publications of the University of Alaska's Cooperative Extension Service helped us considerably in our research. In the section titled "Berries, Low and High," we are pleased to present recipes taken with permission from the University of Alaska Cooperative Extension Service publication *Collecting and Using Alaska's Wild Berries*. This excellent publication was the most complete reference to gathering and preparing Alaskan berries that we found, and we are very happy to be able to reprint some of its tasty dishes in our book. Mr. Walter Them, the University of Alaska Cooperative Extension Service agent in Kodiak, has been a supportive and encouraging friend. We appreciate his taking time to look over our manuscript and offer suggestions. We also appreciate his contacting Mr. James A. Smith, editor of the Cooperative Extension publications, and explaining our project to him, thereby gaining us permission to reprint the berry recipes.

A special salute goes to Gretchen Bersch of The Northern Institute in Anchorage. Gretchen, an incredible high-energy lady, made us manuscript copies, rounded up reams of paper to use with the word processor, introduced us to valuable resource people, and got us going in the right direction when the manuscript was written.

And to the Ouzinkie School staff and especially Chuck Coons, principal, goes our appreciation for their helpfulness and their typewriters. Chuck even set me up with my own desk for awhile, in the storage closet across from his office. We are also very grateful to Yukon Office Supply in Kodiak for letting us type the final copy on their memory typewriter.

When we had the completed manuscript in hand and were somewhat at a loss as to what to do next, Roger Page stepped in and helped us plan a preliminary layout. His expertise was a valuable asset, and we only hope he is forgiving about being awakened too early in the morning with questions about a book on weeds.

Some extra-special touches were added to our book with a little help from our friends. When I asked Kodiak artist Janet Quaccia if she could do a drawing that would capture the essence of Spruce Island, she drew the picture that appears facing our title page. We think she achieved her purpose admirably. Janey Wing Kenyon contributed her lettering talents to enhance our preliminary design. Daniel Konigsberg took class pictures, and Teddy Panamarioff helped with the preliminary paste-up when we were down to the wire and needing extra help. We are proud to have their contributions to our island herbal.

Thanks also go to friends who have given suggestions and words of encouragement. An extra tip of our collective hat goes to Cathy Ray Klinkert, who researched part of the berries section, even though she is in no way involved in our class other than as a well-wisher who is looking forward to the completed publication. And to the Kodiak Community College staff, who have accepted collect telephone calls, answered letters, donated supplies, and given continuous support and encouragement during the birthing process of this book, we send our Gold Star Award.

A considerable amount of research has been done by members of the Plants Class to put this book together. Some of it has been done as classwork in the course of the last three years. Some of it has been done specifically for the purpose of adding to the material presented in this book. Every Plants Class member who has been involved in the book-writing project has in some way gone beyond the ordinary expectations for participants in a once-a-week adult education class session. These people have demonstrated that some wonderful things can happen in an adult education class. They have become so interested in learning and sharing their knowledge that they have given freely of their time and energy with no thought of gain. They have enjoyed themselves in the process. And they have acquired new and useful information.

They have read books, written reports, interviewed local people, written and tried recipes, walked over miles of woodland and bog to collect plant specimens, and typed pages of material. In short, they have been wonderful!

One of the last class assignments of this year was to submit a written essay entitled "What is an Herb?" It seems appropriate to end this preface and get on with the book by sharing parts of these essays with all who read this book. So here they are, compliments of the Ouzinkie Botanical Society, also known as Plants Class.

My Idea of an Herb

An herb to me can be a tree, shrub, vine, lichen, or other member of the plant world, be it oriented to water, rock, or land. It is a plant that has some beneficial use to man and animal. Perhaps it is a medicine, a dye, or a food. Perhaps it tastes good or smells good. Perhaps it cleanses our system or improves our sight or rids us of pests that are injurious to our health. Perhaps it gives us nutrients we need. An herb contributes in some way to our health or well-being.

Rosemary Squartsoff

What is an Herb?

An herb is Mother Nature's own medicine cabinet. It soothes, promotes healing, helps the human body out in various ways to heal minor ailments that have plagued mankind since the dawn of his beginnings.

When properly prepared its uses for curing are practically endless.

Herbs are natural and healthy for you. Herbs were around for many centuries before aspirin was invented. Herbs don't contain any harmful manmade chemicals found in expensive medicine that sometimes are foreign to our body's chemistry. Natural chemicals found in herbs tend to match our natural body chemistry with a common bond between plants and man.

Greg Wolfer

What is an Herb to Me?

Herbs to me are a precious commodity — leaf, stem, flower, and seed are each invaluable. Herbs are a gift from God. From ancient times they have provided natural foods and medicines. There are herbs for every ailment; herbs to calm the spirit; even herbs for love!

Herbs! I could write a book on the topic of herbs — the wonders, the miracles of herbs — not to mention their many uses and the many recipes for cooking them. And yet millions of Americans, including Alaskans, do not utilize these plants.

Politically speaking, were herbs put on the "endangered species list," as are the bald eagles and the whales, and it became a necessity to preserve them, they would have no stronger advocate than I.

Angeline Anderson

What is an Herb?

An herb is any plant growing above the ground, with stems and leaves that wither at the end of the summer or growing period. Therefore, this would exclude shrubs and trees as their stems do not die but live from year to year.

Herbs have been around since the earth's early beginnings. Man has eaten them since he first walked this earth. Bitter herbs were food associated with the Jewish Passover holiday.

Long ago, herbs were used for curing and preventing most diseases and anything that ails us.

Today, herbs can still be used for medicinal purposes if we only know which one to use for a given situation. A good cook uses herbs for seasoning anything from bread to main dishes. With so many kinds to choose from, he can be creative. Also, herbs are made or brewed into teas for a pleasant and stimulating drink.

Georgia Smith

Here's to Herbs!

I've always liked the saying, "A weed is an herb for which a use has not yet been discovered." Perhaps the Ouzinkie Botanical Society should restate it to say, "An herb is a weed that we learn how to use in Plants Class."

We certainly *have* learned — and we've done it together. Finding out the many things an herb can be has been a good and gratifying experience. And so, my salute to you, mysterious herb — the pleasure has been all ours. May the search continue!

Fran Kelso
Ouzinkie, Alaska
1983

P.S. This book may not be our last word; we are still learning about plants. For example, Rosemary uses a plant now that her husband, Fred, taught her about. The Aleut people call it "wild carrot." We don't know the botanical name of this wild vegetable yet. When we do, we'll let you know.

We are also learning about mushrooms and sea vegetables. As we discover more, we may wish to write the information and pass it along, in the form of supplements to this volume.

INTRODUCTION: USING THIS BOOK

Plants discussed in this book are divided into six categories. The biggest category contains most of the smaller plants and is called "A Gathering of Herbs." Other categories are "Ferns," "Berries, Low and High," "Shrubs," "Trees," and "Poisonous Plants." The poisonous plants are included for identification purposes; they do not have culinary or medicinal uses.

Each plant in the six categories is catalogued in the following manner:

1) Family name (next to the page number)
2) Common names, both local and from other areas
3) Latin name (genus and species)
4) Aleut and Russian names, if known.

For example, the cultivated garden rose has some wild relatives that grow in Alaska. The Latin name, genus and species, of the wild rose found in the Kodiak area is *Rosa nutkana*. The Russian-Aleut natives in the area call this shrub "shipóynik."

Common names are listed with the name most familiar to us, on Spruce Island near Kodiak, first. Most plants have many common names; even on our island a plant may be known by more than one name. After the first common name, the other names are listed in order of frequency of use with the most obscure name last.

For each plant presented in the six categories, we have written a description of the genus and species found in the Kodiak area, and the habitat in which the plant normally grows. We have kept the language as simple and non-technical as possible in the hope that other interested adults, no matter what their educational level, will be able to use the material easily. When it has been necessary to use scientific or medical terms to describe the plant or its properties in some way, we have included these terms and their definitions in a glossary at the back of the book.

An additional note on genus and species: Alaskans living in areas of the state other than Kodiak may quite possibly have one or more other species of the same genus growing in their area. Perhaps their plant species will be slightly different in appearance. However, general characteristics of the genus will remain the same. In many cases we have listed ways in which Alaskans outside our area have used the plants discussed in this book. In some cases — where a different species in another area of the state has an important use — we have noted that use and listed the appropriate Latin name. For example, our relative of the Eskimo potato has edible leaves, but has little, slender roots instead of small tubers that can be boiled and eaten as a vegetable. In Latin, the plant in our area is *Claytonia sibirica* (Alaska spring beauty), while further north, the varieties with the edible roots are *C. acutifolia* or *C. tuberosa*.

RUSSIAN AND ALEUT PLANT NAMES

Some of the foraging done to complete this book has been for information rather than for local flora. One of my last tasks before typing our final manuscript was to get a phonetic English spelling for the Russian plant names we had learned.

I had informative sessions with Marian Brown and with Nina Gilbreath in Kodiak and compared their spellings and pronunciations with those we had adapted. Marian Johnson of the Kodiak Historical Society also furnished us with valuable information she had gathered. With these three sources, we had a fairly accurate listing of the Russian plant names we knew.

I wished to take this information to Pleasant Harbor to Father Gerasim, who is quite proficient with the Russian language, to get a final check for accuracy and to pick the best of alternate spellings. Father Gerasim was working to build, on a towering hill, the New Valaam Russian Orthodox Monastery. Since Spruce Island, once the home of Saint Herman of Alaska, is considered by many to be a holy place, Father Gerasim and other monks had come to the island hoping to fulfill Saint Herman's wish that the monastic tradition be continued there.

Time was getting short. It was early February, 1984. Only one weekend remained before I was to return to Kodiak to type the manuscript one last time.

Unfortunately, that Friday and Saturday we had the heaviest single snowfall I had seen since building my cabin in 1977. The three-mile trail to Pleasant Harbor, following a wide clear-cut area where a road was once planned, would be plugged to the brim with snow. Nevertheless, when the sun came out Sunday morning, I persuaded my good friend, Teddy Panamarioff, to accompany me and we set off from Banjo Beach.

The hike was a comedy. The snow was 3 to 4 feet deep in much of the clear-cut area, and traveling was at a snail's pace. A great deal of determination and a touch of insanity was undoubtedly what got us there at last. A hike that normally would take us forty-five minutes to an hour took nearly three hours to complete. But the laughter that accompanied our efforts at thrashing through the snow made it all worthwhile. And when Peter Chichenoff appeared while we were at work on the plant names with Father Gerasim, we were really astonished. He had walked all the way from Ouzinkie, a mile and a half farther away from Pleasant Harbor than my cabin at Banjo Beach.

When the work on the plants was completed, we stayed for church service and supper at the monks' house. Then the three of us waded home to Banjo Beach together in the dark. We had left at 11 a.m. and we got back at 10 p.m. It was a beautiful day for us all.

While comparing notes with our language advisors, we learned some interesting things. First, much of this old herbal knowledge was handed down verbally, rather than in writing. As a result, both Russian and Aleut words vary a little from village to village. To confuse matters further, we found a few plant names that appeared to be a Russian word and an Aleut word mixed together. The word as we present it in this book is the one most commonly agreed upon among the sources we consulted.

Both Nina Gilbreath and Father Gerasim looked up the Russian spelling of the plant names first. Often they would find the Russian root word that was the closest to the pronunciation we knew, and from these they supplied us with a noun form and an English spelling. Many times these roots would have a meaning that described some characteristic of the plant. For example, "zhólti golóvnik," or goldenrod, translates as "yellow-headed." Yarrow, or "poléznaya travá," means "the healing herb" in the Russian language.

It is much more difficult to provide proper spellings for Aleut plant names, as very few people know written Aleut. Whenever possible, we used spellings found in Jeff Leer's *A Conversational Dictionary of Kodiak Alutiiq*. The remaining Aleut names we spelled with inadequate English phonetics. There is always more to learn.

COLLECTING PLANTS

Two cautions about collecting wild plants: First, *never* use a plant for food or medicine unless you are sure you have properly identified the plant. And secondly, when gathering wild plants, *never* take all the plants in the area. Leave some there to repopulate the species, so that you and others may have the pleasure of using the plant again.

In his *Root, Stem and Leaf*, Glen Ray lists nine techniques to use when gathering plants. These techniques are designed to help preserve the natural environment while still using the resources available. They are part of a system called "traditional conservation." We feel these techniques are important for Alaskan foragers to follow, so have repeated them here:

1) Learn the habitat and conditions under which each plant flourishes.
2) Know the area in which you live well enough to know where each plant can be abundantly found.
3) Take time to ask Native elders if the locale where you would like to harvest a plant is not already a harvesting spot for a group of people.
4) Find a place to harvest not already harvested.
5) If the plant seems not to be abundant in the area where it is found, it would be best not to harvest until it can be found growing abundantly. If one feels that some harvesting is possible then take only a few plants or only some portion of several plants.
6) Leave the roots of perennials intact along with a portion of the leaves so the plant can regenerate.
7) Take only a part of a plant so the plant can flower and reproduce.
8) Take only what can be processed and used.
9) Take time to enjoy the process and appreciate the surroundings.

PLANTS AS FOOD

Where applicable, we have included a note about the nutritional value of each plant. Scientists have not tested wild edibles as thoroughly as they have domestic crops; consequently, nutritional information is not always available. However, the testing that has been done has shown that edible wild plants are consistently higher in nutritional value than their domestic cousins. Our bodily well-being can only be improved by seeking out these free foods and adding them to our diet.

Wild food foragers should keep in mind that fresh edibles of any kind begin to lose their nutritional value as soon as they are harvested. Therefore, eat wild foods as soon as possible after they are gathered to gain the fullest nutritional benefit they can provide.

For each edible plant, complete instructions have been given for preparing it. In many cases, specific recipes are listed for the plant under discussion. We have tried many of these recipes at our potlucks and in and out of class. Some of them we have written ourselves. We offer them with the hope that you, the reader, will discover for yourself how very tasty these wild dishes can be. We suggest, too, that you experiment with these recipes and modify them to suit your own taste, if necessary.

PLANTS AS MEDICINE

As we learned ways to prepare our wild plants for the dinner table, we also learned ways they have been used for medicine. As one Russian-Aleut grandmother told us, "there's medicine growing all over out there in the woods." She said that in the days when her mother and grandmother were young, there were no doctors like we have today. Instead, there were Aleut doctors who had learned of the medicinal uses of the plants around them and had gathered and stored them in case they were needed in the winter months. When a person was sick, the Native doctor treated the patient. He mixed the herbs according to his knowledge of their uses. The patient was usually cured, said the grandmother, though sometimes the process took longer than it does these days, because there were no wonder drugs or pills for everything. But people did get well.

One reason people got well is probably the high nutritive value of many of the wild plants used. Often a disease was contracted because of a vitamin deficiency that allowed the sickness to gain a hold. It is quite possible the herbs prescribed often contained the right vitamin content to fill the body's need and eventually restore it to health. An obvious example was the use of wild greens high in vitamin C by early sea-going peoples to counteract the effects of scurvy. Though other curative properties existed in the herbs as well, the high vitamin content in many of them certainly contributed to their usefulness.

We are including some of these old medicinal uses to further the reader's knowledge about these plants, and not with the suggestion that these teas and tonics be substituted for medicine provided by a physician. Nor is it a good idea to use herbal remedies and patent medicines at the same time, unless the user first consults his doctor. Harmful side effects or overdoses might occur.

Neither do we advocate herbal remedies in place of a doctor's care for a very sick person. We can provide no guarantees of the effectiveness of many of these natural cures; anyone wishing to try them must do so at his own risk. However, for a simple ailment, perhaps these mixtures prepared from wild plants can provide relief and save a person some discomfort, or be helpful if a person is ill and cannot immediately reach a doctor for aid.

SPECIAL RECIPE AND INTERVIEW CHAPTERS

We have included some special sections, in addition to specific plant information, in our publication. One is a short section on tea recipes, as prepared by Rosemary Squartsoff and Plants Class. Another is a small collection of natural dye recipes. In addition, we have written down our favorite regional fish and game recipes and those of some of our friends, so a person might conceivably prepare a whole meal using recipes found in this book. And, last but quite important, we have included an interview with two of our favorite Ouzinkie babushkas (grandmothers) about the ways they remember local herbs being used.

A GATHERING OF HERBS

BUCKWHEAT FAMILY

MOUNTAIN SORREL, Sourgrass
Oxyria digyna

DESCRIPTION: Mountain sorrel, a perennial, grows from a fleshy taproot. Its erect stems reach from 4 inches to 2 feet tall; there are 1 to 2 kidney-shaped leaves on each stem. The small flowers which cluster at the top of the stalk are greenish to crimson.

HABITAT: Mountain sorrel is found at sea level. It likes moist, rocky, gravelly spots, such as sheltered gulches.

EDIBLE PARTS AND NUTRITIONAL VALUE: The leaves (but not the roots) can be eaten raw or used as a potherb (prepared like spinach).

This plant is a good source of vitamin C. However, because the mild sour taste comes from oxalic acid present in the plant, mountain sorrel should not be eaten often and in great quantities (see additional information under sourdock, following).

. . . Mountain Sorrel

WAYS TO PREPARE FOR EATING: One of the tastiest wild greens, mountain sorrel leaves make a good lettuce substitute in sandwiches. They can also be included in salads and soups. Try them wilted with vinegar and bacon, or steam the tender leaves briefly; don't overcook them. They make tasty creamed soup or purée.

The Eskimos ferment the leaves into a kind of sauerkraut.

Mountain Sorrel and Fish Soup
Bradford Angier,
Feasting Free on Wild Edibles

Make a fish stock of heads, tails, bones and fins (try with salmon and white fish both): bring to a boil in cold water to cover, then simmer for 1 hour or longer with 1 onion (chopped) and 1 clove garlic.

Just before eating, heat, for each diner, 1 cup strained fish stock, ¼ teaspoon salt, and black pepper to taste. Stir in 1 cup mountain sorrel (or sourdock) purée and simmer 5 minutes; serve hot with a pat of butter on top.

SOURDOCK, Arctic Dock, Dock, Curly Dock, Wild Spinach, Wild Rhubarb
Rumex
Kislítsa (Russian)

DESCRIPTION: Several species of *Rumex* grow in Alaska. All are edible. Each of these perennial plants has a stem, which can get to be 3 or 4 feet tall, sprouting from its center. The stem grows from long, yellowish roots. Sometimes the roots resemble fat carrots, except they grow sideways instead of up and down. The leaves are long, arrowhead-shaped, with wavy or curly edges. Usually dark green, sometimes they are reddish close to the stalk. Most of the leaves grow close to the ground, with a few leaves climbing the stalk in an alternate pattern. The flowers, clustered at the top of the stalk, form tiny, thin-winged reddish seeds that are released on windy days.

Sourdock is called "wild rhubarb" by people in the Kodiak area. Another wild rhubarb, *Polygonum alaskanum*, does not grow in our area, though both plants are in the buckwheat family. The *Polygonum* plant is used in much the same way as domesticated rhubarb.

One *Rumex* species, *R. acetosella*, has rather small leaves and, though edible, is more commonly used for making dye. It is more branching and delicate-appearing than sourdock.

HABITAT: Grows in wet meadows, fields, along roads and slopes, drainage ditches, vacant lots.

EDIBLE PARTS AND NUTRITIONAL VALUE: Stems, leaves, seeds and roots can be eaten.

This plant has more vitamin C than oranges and more vitamin A than carrots. It contains calcium, iron, potassium, phosphorous, thiamine, niacin and riboflavin.

Sourdock contains oxalic acid, which can be dangerous if consumed in large quantities. However, many other domesticated vegetables also contain oxalic acid, and are not harmful if eaten in normal amounts. As long as it is used in moderation, sourdock is a healthful and beneficial wild vegetable.

MEDICINAL USES: Both the leaves and roots of sourdock can be used for medicinal purposes.

Leaves: Apply as a dressing for blisters, burns and scalds. Rub the leaves on the skin to remove the sting of the nettle plant. Squeeze the juice from fresh leaves and put on ringworm or other parasites and fungi. Old-timers recall that leaves of sourdock were

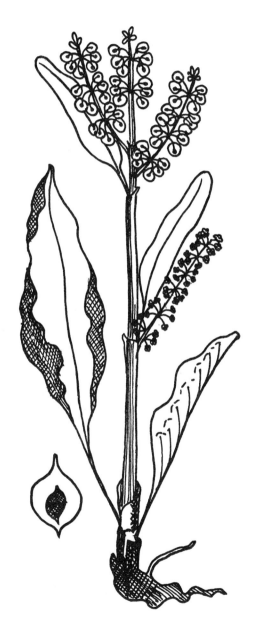

12 BUCKWHEAT FAMILY

. . . Sourdock

eaten in large quantities to purge the system and clean the blood.

Roots: Boil or soak in hot water. Drink very strong to bring on vomiting. Brew a weaker mixture for a gentle tonic, astringent, and laxative. People have taken this tea for stomach and bladder trouble, tuberculosis, and for relief from a hangover. The roots can be dug in the fall and dried for tea in winter. For an ointment for skin problems, boil the root in vinegar, then mix the softened pulp with lard or petroleum jelly.

OTHER USES: The root can be used to make dye ranging from tan to deep gold.

WAYS TO PREPARE FOR EATING: Leaves and stems can be eaten raw or boiled. Prepare the leaves like spinach and the stems like rhubarb. The raw leaves, with their pleasant lemony flavor, are very tasty in salads; they also make a good lettuce substitute in sandwiches.

We suggest steaming young leaves, cooking only until tender. They take just a short time to cook. If left in the pot too long sourdock becomes quite mushy. Young sourdock leaves are also good wilted with vinegar and bacon.

When cooking older plants, bring to a boil, drain water away, and add more water. Cook just until tender.

Sourdock can be used in creamed soups and purees.

The Eskimos cooked, chopped, and mixed sourdock leaves with other greens, then stored them in kegs in a cool place for later use.

The seeds can be ground and made into cakes or cereal.

For additional sourdock recipes, see Mountain Sorrel and Fish Soup (mountain sorrel under Buckwheat Family), Green Noodles from Mars (dandelion under Composite Family), Makrétzi Soup (chickweed under Pink Family), Spruce Island Weed Salad (seabeach sandwort under Pink Family), Early Spring Salad (goosetongue under Plantain Family), and Wildwood Fritters (goosetongue under Plantain Family).

Clam Soup
Bradford Angier,
Feasting Free on Wild Edibles

1 onion, chopped
4 tablespoons butter
2 cups finely shredded young sourdock leaves
2 cups cleaned clam meat, fresh or canned
2 cups rich milk
Pepper

Brown onion in the butter until soft tan. Add sourdock leaves and stir for about 60 seconds — until leaves wilt. Add clam meat and milk. Bring to a slow simmer and cook for no longer than 1 minute, as cooking toughens clams. Sprinkle with pepper to taste and serve.

Creamed Sourdock

2 cups sourdock (can be mixed with other greens)
½ cup rich milk
1 tablespoon flour
Salt
Pepper

Cook sourdock, then drain and chop; set aside. Combine milk and flour, stirring thoroughly. Add sourdock. Stir and cook until slightly thickened. Add salt and pepper to taste. Serve.

BUCKWHEAT FAMILY

Boiled Wild Greens
Rosemary Squartsoff

¼ pound sliced bacon
3 cups water
2 cups sourdock leaves
½ cup petrúshki leaves and stems
1 cup nettle leaves
Salt

Cut bacon slices into squares and fry until semitransparent. Add water; bring to boil. Chop the sourdock, petrúshki and nettles. (All of these should be fresh and tender.) Cook in the boiling water 5 to 15 minutes. Salt to taste. Serve.

CLUB MOSS FAMILY

FIR CLUB MOSS, Club Moss, Spike Moss, Christmas Greens
Lycopodium selago

DESCRIPTION: Fir club moss is a low perennial plant that has been around since prehistoric times, when some types grew to the size of trees. Much coal was formed from this plant. It is not a true moss, but is closer to a fern. Its branches stick up from the stem with leaves that look like spruce needles. In the summer, some of the branches have spikelike ends that look like miniature clubs.

WARNING: All parts of this plant — except the spores — are poisonous when eaten. Fir club moss is safe when used externally.

HABITAT: Fir club moss likes acid soil, and grows in the woods and grasslands.

MEDICINAL USES: This plant is poisonous when eaten. Only its spores should be taken internally.

The whole plant can be boiled in water for an eyewash. A piece of club moss placed on one's head is said to get rid of a headache.

A powder made out of the spores is supposed to stop nosebleeds and bleeding from a wound. This powder has also been helpful for absorbing fluids from damaged tissues in different kinds of injuries. In the past, pills were coated with this spore powder to prevent them from sticking together. Also, the spore powder can serve as a dusting powder for various skin diseases. Recent research indicates it may be helpful in treating herpes.

Fir club moss spores have been used to treat uterine problems, knee problems, swollen thighs, water retention, and dropsy (excessive fluid collected in the body).

OTHER USES: Fir club moss is also called Christmas greens because it stays green all winter and can therefore be picked for Christmas decorations.

The spores of these plants flash with a hissing noise when ignited. They are used in the manufacture of fireworks.

This plant was probably the "cloth of gold" of the ancient Druids, as it was supposed to provide protection against black magic.

COMPOSITE FAMILY

COMPOSITE FAMILY

CARIBOU LEAVES, Stinkweed, Silverleaf, Wormwood
Artemisia Tilesii
Polín (Russian) — Noun form
Políbnya (Russian) — Adjective form

DESCRIPTION: Polín, as it is called in the Kodiak area, is a perennial plant that grows on a single stem, up to 2 or 3 feet tall. Its leaves, which grow close to the stem, are green on top and silver-green underneath. They are shaped something like the foot of a raven: narrow, with 3 to 5 long "toes." These leaves have tiny hairs on both the bottom and the top. In late summer, clusters of small flowers grow in a spike at the top of the stem. The plant is most easily recognized by the pleasant smell it gives off if brushed against: pungent and rather sweet. "Almost like spearmint," says Sasha Smith.

This is one of our "pet plants." For three years we knew only its Russian name. We were told of its uses and even collected a sample, which my cat destroyed before we could get it identified. A short time ago, Angeline Anderson discovered a picture of polín and an article titled "Medicinal Leaves of the Tahltans" in the April, 1983 *ALASKA®* magazine. Local sources and a sample of the dried plant collected by a Ouzinkie friend confirmed that we'd finally discovered the plant's Latin name and a British Columbian name, "caribou leaves." We had also collected an article called "To Heal the People," from *Northwest Arctic Nuna*, in which the Inupiaq healer, Puyuk, describes the same plant, calling it "stinkweed." Priscilla Kari describes its uses in her *Tanaina Plantlore*, and notes its similarity to wormwood, a plant of the *Artemisia* genus long known to herbalists.

HABITAT: In our area, polín is usually found on the mountainside in sunny places and in a few garden spots where it has been transplanted by local people. In Kodiak, Sandra Coen discovered it flourishing along creek banks in the Buskin River area.

MEDICINAL USES: Tahltans and other people of their area in British Columbia take polín (caribou leaves) as a tea for colds, as a gargle for sore throats, or as a wash for sore eyes or cuts. They also drink the tea to relieve constipation, stop internal bleeding, and alleviate kidney problems. People in different parts of Alaska have used the plant for similar purposes. For a tea, the healer Puyuk recommends boiling the leaves for 15 to 20 minutes in water (1 ounce leaves to 1 pint water is a fairly standard herbalist's measure), straining out the leaves, cooling the liquid in a non-metal container, and then drinking. The taste is bitter, but Puyuk says the tea is a "good medicine." However, use this tea sparingly. A large dose could upset your system.

Locally, the fresh leaves are steeped in boiling water and the tea taken in small doses as a blood-purifying tonic. It is said to help blood circulation and to dissolve lumpy varicose veins. A friend here told us of a relative of hers in another village who drank polínya tea for this purpose, and it helped the varicose veins go away.

Another use we learned in Ouzinkie was healing old cuts that were infected and wouldn't heal. Steep the dried or fresh leaves in boiling water for a few minutes, then dab the infected area with wet leaves. It will sting, but if the process is continued regularly the infection will be stopped and the wound healed. We were told about a small boy in Ouzinkie who had sores on his face that wouldn't heal. The polínya leaves were tried, and the sores went away.

Another local lady tells us she learned to chew the leaves for relief from various ailments, including

. . . Caribou leaves
colds and flu. I've tried chewing the leaves — they are strong-tasting but not unpleasant, and remind me of eucalyptus. Puyuk also recommends chewing polín leaves or "stinkweed" for colds, flus, fever, headaches, and ulcers. She says chewing the leaves also helps insure a healthy pregnancy.

Julia Pestrikoff in Port Lions says the seeds from polín used to be a remedy for heart troubles. She called it silverleaf.

Another of our local use consultants, Jenny Chernikoff, was taught to make a poultice of polín as a help for arthritis pains. The leaves are heated and put over the place that is hurting.

DANDELION
Taraxacum
Odoovánchik (Russian)

DESCRIPTION: Dandelion can be a perennial or biennial. The stem is erect, and the plant often has many flowers. The leaves grow in a rosette at the base of the plant. They are long and narrow with deep teeth. The name of the plant comes from French, "dent de lion" (teeth of the lion), because of the shape of the leaves and of the yellow petals. The bright yellow flowers grow in a head. These flowers are replaced by a puffy seed pod. A milky juice comes from the hollow flower stem when it is broken. The root of the plant looks something like a carrot.

HABITAT: Dandelions grow in fields, along roadsides, and in your lawn.

EDIBLE PARTS AND NUTRITIONAL VALUE: Roots, crowns (parts between roots and ground surface) and tops — from young leaves to flower buds — can be eaten.

The fresh leaves are an excellent source of vitamins B, C and A, and of calcium, potassium, phosphorous and sodium. (Plant is especially high in calcium.)

MEDICINAL USES: Dandelion has two very important uses:

1) It helps cause the formation of bile.
2) It removes excess water from the body.

To remove poisons from the body, provide a tonic or stimulant, or supply a mild laxative, it is best to drink freshly squeezed juice.

For rheumatism, gout, and stiff joints, try the following cure for 8 weeks:

1) Add 2 tablespoons fresh roots and leaves to ½ cup water.
2) Boil, then steep 15 minutes.
3) Take ½ cup, morning and night.
4) Also take 1 to 2 glasses of water each day, with 3 tablespoons of juice pressed from the roots and leaves per glass. Eat some fresh greens daily in salad.

For colds, prepare a tea in the following manner: Add 2 teaspoons, packed, of fresh, chopped greens per 1 cup of boiling water. Let stand 8 hours and drink 1 cup at a time.

WAYS TO PREPARE FOR EATING:

Roots: Scrape, slice, and boil the roots in salted water until just tender when pierced with a fork, then serve as a vegetable. Dandelion is related to chicory, and, like its relative, its roots are suitable for a coffee stretcher or substitute.

Crowns: Boil with the leaves or alone, as a vegetable.

Flowers: Make dandelion wine, or put in pancakes or fritters.

Leaves and Buds: Prepare as greens; can be eaten raw or cooked (good wilted with vinegar and

. . . Dandelion

bacon). If older leaves are bitter, drain well after they reach a boil, cover with fresh water, and bring to a second boil. Or, pick leaves late in the year — the first frost kills the bitter taste. Hang larger plants by the roots to dry; crumble and use like parsley.

For additional dandelion recipes, see Clover-Bright Salad (clover under Pea Family), Early Spring Salad (goosetongue under Plantain Family), and Spring Beauty Salad (Alaska spring beauty under Purslane Family).

Summer Salad

A combination of dandelion leaves, fireweed, and lambsquarter can be added to head lettuce. Add cooked, diced beets or pickled beets. Arrange sliced hard-boiled eggs on top; serve with French dressing.

Banjo Beach Omelette
Fran Kelso

2 or 3 eggs, beaten
⅓ cup milk
½ cup (packed) cleaned and stemmed young dandelion leaves
2 slices bacon
Dash pepper
2 to 3 tablespoons butter or margarine
⅔ cup grated cheese

Mix beaten eggs and milk. Chop dandelion leaves and mix with eggs and milk. Fry bacon until crisp; crumble into egg mixture. Add pepper.

Melt enough butter in a medium frying pan to allow eggs to cook slowly without burning. Pour egg mixture into hot butter, spreading ingredients evenly in the pan. Sprinkle grated cheese over top and cover, cooking slowly until eggs begin to firm.

As omelette cooks, lift edges to allow liquid on top to run underneath. When the eggs are cooked through, loosen from pan with a spatula and flip half on top to form a half-circle. Remove to plate and serve . . . a meal in itself.

Variations: Try adding small chunks of cream cheese, tomatoes, a handful of sprouts, or all three. Or include any of your other favorite omelette ingredients. Other greens can be substituted for the dandelion leaves.

Green Noodles from Mars
Reprinted by permission © *The Herb Quarterly*, Newfane, Vermont 05345

4 ounces (2 cups, packed) finely chopped dandelion greens
2 eggs
1 teaspoon salt
1 to 1½ cups flour

Put dandelion greens and eggs in blender and whirl until smooth. Place in bowl, add salt, and beat in flour until dough is very stiff. (Adjust amount of flour to the moisture of the greens.) Turn dough out onto a floured surface and knead about 5 minutes. Roll dough with floured rolling pin until it is a noodle-thin sheet. Let stand for about an hour to dry. Roll up in jellyroll shape; cut in strips and shake out. For soup, drop in boiling stock about 15 minutes before serving. For cooked noodles, put in boiling water for about 8 minutes.

Variations: Sourdock or lambsquarter could be used instead of the dandelion greens.

Braised Dandelion Greens
Sandra Coen

1 pound (approximately 10 cups) dandelion greens
1 medium onion, sliced
1 large apple, sliced
2 tablespoons oil
½ teaspoon salt
⅛ teaspoon pepper
2 to 3 tablespoons cider vinegar (optional)

Chop dandelion greens, including ribs of leaves, into 1-inch pieces. Sauté onion and apple slices in oil for a few minutes. Add chopped greens, salt and pepper; stir-fry slowly 15 minutes. If desired, add cider vinegar and cook 5 more minutes.

Dandelion Coffee Substitute

3 cups fresh dandelion roots
1 quart water
1 tablespoon cold water
Mild honey

Preheat oven to 275 degrees.
Wash roots under warm running water. Spread on baking sheet and place in the warm oven until roots are brown as a nut all the way through (check after 10 minutes).
For each quart of brew, use coffee grinder to prepare 2 tablespoons grounds. Heat the quart of water in a pan until almost boiling, then reduce heat — do not allow to boil. Pour ground roots onto hot water. Keep pan over very low heat for 10 minutes. Remove pan from heat; allow ground roots to settle. Add the tablespoon of cold water. Drip brew from pan without disturbing grounds. Serve with honey.

Dandelion Cheese Casserole
Brenda Theyers-Wilson

1 onion, chopped
½ pound cottage cheese
½ pound feta cheese
3 eggs
Dill
Salt
Pepper
4 tablespoons butter, melted
Several cups chopped dandelions

Mix together the above ingredients. Make a batter of the following:

1½ cups water
3 eggs
1½ cups flour

Pour half the batter into an 8 by 8-inch dish. Top with dandelion mixture, then pour on other half of batter. Bake at 350 degrees for 1 hour.

Elizabeth Insley's Dandelion Recipe

3 cups dandelion leaves
2 tablespoons butter
3 slices bacon, cut in small pieces
Salt
Pepper
1 egg
¼ cup vinegar (more to taste)
1 hard-cooked egg

Wash dandelion leaves well; let drip, but do not dry. Crisp the bacon pieces. Melt butter, add dandelion leaves and bacon, and cook, turning often, until leaves are wilted. Add salt and pepper to taste. Beat together egg and vinegar and add to cooked greens. Slice hard-boiled egg over top.

Dandelion Blossom Pie
Jim Woodruff

Pick enough dandelion blossoms to fill a 3-quart saucepan ¾ full. Wash blossoms thoroughly and cover with water. Cook 45 to 60 minutes, until blossoms are tender and water is permeated with dandelion flavor. Separate blossoms from liquid. Measure the following:

2 cups dandelion liquid
1 cup sugar
3 tablespoons quick-cooking tapioca
2 cups (packed) cooked dandelion blossoms
½ teaspoon nutmeg

Cook liquid, sugar, and tapioca until tapioca is transparent. Fold in the blossoms. Add nutmeg. Place in an 8-inch graham cracker crust or baked pie shell. Garnish with whipped topping and serve.

Dandelion Wine
Fran Kelso

15 quarts dandelion blossoms
3 gallons cold water
15 pounds sugar
1 yeast cake or about ½ ounce yeast
Rinds and juice of 1 dozen oranges
Rinds and juice of 6 lemons
2½ pounds raisins

Place blossoms in cold water and simmer for 3 hours, then strain the liquid and mix it with sugar.

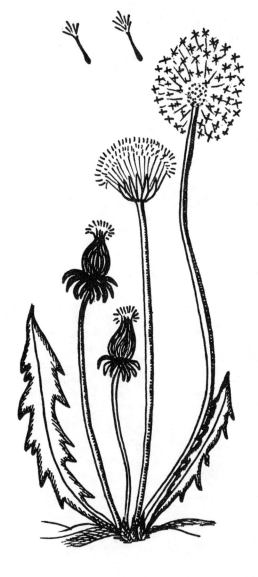

Bring to a boil; strain through cheesecloth. When lukewarm, add yeast. Let mixture stand for 2 or 3 days, skimming it each day.

Simmer thinly peeled rinds of oranges and lemons in a little water for ½ hour. Add cooked rinds and orange and lemon juice to yeast mixture. There should be 5 gallons. Pour into a 5-gallon cask or crockery container, and add raisins. Leave cask open for 1 day, then seal it tightly and let it stand for 6 months before bottling. The wine improves with aging.

GOLDENROD
Solidago multiradiata
Solidago lepida
Zhólti golóvnik (Russian)

DESCRIPTION: Goldenrod is a perennial which can grow up to 3 feet tall. Its stem has only a few branches close to the main stalk, or, sometimes, only a single stem. Goldenrod leaves are alternate, smaller toward the top of the plant. Golden-yellow flowers cluster on the top section of the stem.

HABITAT: Likes dry, open areas, such as treeless hillsides.

EDIBLE PARTS: Dried leaves and flowers, either alone or blended with other ingredients, make an aromatic tea. Goldenrod can be mixed with other medicinal herbs to help improve their flavor.

COMPOSITE FAMILY

. . . Goldenrod

MEDICINAL USES: Leaves and flowers can be steeped in a tea which, taken cold, acts as a stimulant and helps get rid of gas. Goldenrod was once taken for kidney problems. As it acts as an astringent, it can be useful for internal hemorrhaging or diarrhea. Infuse 1 ounce flowering tops in ½ cup water, then steep 5 minutes. Drink 1 cup daily, made fresh each time.

Fresh leaves and flowering tops can be crushed and placed on cuts, sores, or insect bites.

OTHER USES: Stems with flowers can be picked for a steambath switch. The flowers can also be boiled to make a yellow-gold dye; boiling the whole plant produces green dye.

PINEAPPLE WEED, Wild Chamomile
Matricaria matricarioides
Aramaaskaag (Aleut)
Romáshka (Russian)
Aromáshka (In Ouzinkie)

DESCRIPTION: A perennial herb, pineapple weed is a low-growing plant, rarely taller than 9 inches. The leaves of the plant are small and finely cut. Unlike those of true chamomile, the flowers have no white petals, but only the yellow centers that resemble conical balls.

The fresh plant smells somewhat like pineapple, which accounts for its name.

HABITAT: Found in open fields and waste places, or untended lawn areas. It likes to grow near people.

EDIBLE PARTS: The whole plant can be used as a tea, the flowers in salads.

MEDICINAL USES: Pineapple weed has the same properties as chamomile. The yellow flower centers are the most important medicinal part, but the leaves are also good as a tea. This plant can soothe nerves, remedy delirium tremens (D.T.s), and prevent nightmares, or can be taken for a general tonic. For a pleasant tea, add 1 ounce of the yellow blossom balls to 1 pint of boiling water. *Always* prepare the tea in a covered vessel, to prevent the escape of steam — the medicinal value of the blossoms is decreased by evaporation. Let the tea steep for at least 10 minutes.

For head and chest colds, boil the blossoms in a covered pan of water. Have the person with the cold bend over the uncovered pan, and place a towel over both the person's head and the pan. Breathing the steam into the lungs will help relieve cold symptoms.

This tea can be helpful as a mild laxative or to relieve nausea. It is also a good eyewash and skin wash.

In our area, the tea has been given to mothers and newborn babies. Boil the whole plant (above-ground portion) in water. Strain. The mother gets a cup; the baby gets a few drops. It acts as a gentle tonic and helps the mother's milk to start.

OTHER USES: Pineapple weed makes a good deodorant. Rub the raw, fresh plant between the hands to remove fish smell.

The tea serves as an excellent rinse for blond hair. Sponge tea over your body for an insect repellent.

For the gardener, pineapple weed is also of value. Spray the tea on new seedlings to prevent damping-off. Plant it in your vegetable garden as a companion plant. Disperse seedlings around the garden to help sickly plants to health.

COMPOSITE FAMILY

YARROW, Milfoil
Achillea borealis
Qangananguaq (Aleut)
Poléznaya travá (Russian)

DESCRIPTION: Yarrow is a hardy perennial with alternate, very feathery, slightly hairy leaves. Their color is a grayish green. Because of the many small parts of its leaves, yarrow is often known as milfoil, or thousand-leaves. The flowers, atop branched stems, are a flattened mass of tiny flowerets, either white or pale lilac.

Yarrow gets its generic name, *Achillea*, from the legend that soldiers in the army of the Greek hero, Achilles, used this plant to heal their wounds during the Trojan War.

HABITAT: Yarrow can be found growing in almost any kind of soil.

PARTS USED: Every part of the plant that grows above the ground can be used. Cut the whole plant when its flowers are in full bloom and dry it rapidly, at 90 to 100 degrees.

MEDICINAL USES: Like dandelion, yarrow is good as a general tonic and a mild laxative; it is also used to help stimulate the flow of bile. The plant can be either boiled or steeped in hot water.

For a good remedy for severe colds, make an infusion of 1 ounce dried yarrow to 1 pint boiling water. Steep 5 minutes. To a wineglass of this tea, add 1 teaspoon honey and 3 drops hot-pepper sauce. Drink this dose 3 times a day. Stay covered after drinking the tea, as it will open the pores and cause heavy sweating.

The same tea is helpful in soothing menstrual cramps and slowing down extra-heavy menstrual flow.

For stuffed-up sinuses, boil the plant in water and breathe the steam.

Yarrow has long been used to help increase appetite, relieve stomach cramps and gas, and aid gallbladder and liver problems. Kodiak people took it to combat asthma. It has also been taken to help stop internal hemorrhaging, especially in the lungs. Locally, the tea has been found to give relief from a hangover.

Water in which yarrow has been boiled makes a good wash for sore eyes, skin irritations, chapped hands, and all kinds of wounds and sores. A hot pack of the cooked or raw wet leaves can be put on an ache or open sore to help healing and stop infection. For a toothache, chew the fresh leaves, or dip leaves in hot water, wrap in a cloth, and apply to the sore spot on the jaw.

OTHER USES: A yarrow infusion can be applied as a hair rinse that is said to prevent baldness. The fresh plant can be rubbed on skin and clothes for a mosquito repellent.

CROWFOOT FAMILY

MARSH MARIGOLD, Cowslip
Caltha palustris asarifolia

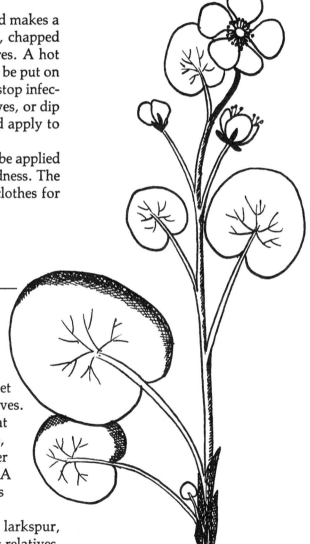

DESCRIPTION: A perennial plant with stout, hollow stems from 1 to 2 feet tall, marsh marigold has only a few leaves. These are large, rounded, and somewhat kidney-shaped, from 2 to 7 inches wide, with scalloped edges. The flowers, either single or in clusters, are bright yellow. A cluster of pods with many seeds follows the flowers.

Buttercup, baneberry, columbine, larkspur, monkshood and anemone are interesting relatives.

. . . Marsh marigold

WARNING: Marsh marigold leaves contain a poison that is destroyed by cooking.

HABITAT: Marsh marigolds are found in marshes and along creek beds. They usually grow in abundance rather than in ones or twos.

EDIBLE PARTS: When gathered in spring before flowers appear, the leaves are tasty. The marsh marigold contains a poison, helleborin, that is destroyed when the plant is cooked. Always boil and drain twice.

WAYS TO PREPARE FOR EATING: Drop leaves into boiling water. As soon as it starts to boil the second time, drain off the liquid. Pour more boiling water over the leaves and drain again as soon as it begins to simmer. This procedure is important; without it the plant is poisonous.

Then steam or simmer leaves in a small amount of water until just tender. Cut into small pieces, add a little salt and a good quantity of butter; a little vinegar can also be added, if desired. Leaves are tasty seasoned with vinegar and bacon.

To cream the leaves, boil them as described above, season with salt, then drain well and chop fine. Make a white sauce in a frying pan with 1 tablespoon butter, 1 tablespoon flour, and salt and pepper to taste. Add ½ cup cream or rich milk and the leaves. Stir well and serve.

EVENING PRIMROSE FAMILY

FIREWEED, Great Willow Herb, Giant Fireweed
Epilobium angustifolium
Cillqaq (Aleut)
Kipráy (Russian)

DESCRIPTION: Fireweed is a perennial plant with a long, single stem from 1½ to 8 feet tall. Its leaves are alternate, long and narrow, smooth on top and a paler shade of green underneath. They grow all along the stem, top to bottom. Large, showy flowers grow in spikes at the end of the stalk. The four-petaled flowers are rose, pink, or white.

HABITAT: Fireweed can be found in meadows, open forests, and sunny hillsides. Look for it also in burnt-over areas and recent clearings.

EDIBLE PARTS AND NUTRITIONAL VALUE: Young stems, leaves, and blossoms are good to eat.

Fireweed is a good source of vitamins C and A.

MEDICINAL USES: Fireweed leaves and stems can be made into tea and enjoyed as a tonic.

To drain a boil or a cut with pus in it, apply a piece of raw fireweed stem. The plant draws the pus out and keeps an infected cut from healing too quickly.

OTHER USES: The down from fireweed seeds has been combined with wool to make blankets, and mixed with cotton or fur to make clothing. It also makes a good fire-starter for campers.

WAYS TO PREPARE FOR EATING: The young stems and leaves can be eaten raw in salads or boiled. They are good with fish eggs. If needed, peel the stems first. It is best to eat young plants, as they tend to get bitter when they grow older.

Very young shoots, when cut in pieces, dropped into boiling salted water, and cooked until they can

. . . Fireweed

be easily pierced by a fork, taste something like asparagus. Once leaves appear the plants may be bitter — bring them to a boil, drain, and boil again. Or, peel the stems of older plants and add them to soups. They will thicken the broth.

Green or dried, the leaves can be used to stretch tea, or a tea can be made from them alone.

For additional recipes using fireweed, see Summer Salad (dandelion under Composite Family), Spruce Island Weed Salad (seabeach sandwort under Pink Family) and Early Spring Salad (goosetongue under Plantain Family).

Marinated Fireweed Shoots
Stacy Studebaker

1 cup or more chopped (1 to 2-inch pieces) fresh fireweed shoots
¾ cup oil
¼ cup vinegar
1 clove garlic, minced
1 tablespoon grated onion
Mint leaves

Place all ingredients in bowl and chill for 2 to 3 hours. Serve.

Fireweed Honey
Betty Blackshear

10 cups sugar
2½ cups water
1 teaspoon alum
18 pink clover blossoms
30 white clover blossoms
18 fireweed blossoms

Put sugar and water in pan; add alum. Place over high heat. Stir. Bring to a rolling boil; boil for 6 minutes. Remove from heat; add blossoms. Strain and pour into hot sterilized jars; seal.

GERANIUM FAMILY

WILD GERANIUM, American Cranebill
Geranium erianthum
Igória (Russian)

DESCRIPTION: Wild geranium is a perennial with leaves that look something like those of the domestic geranium. It grows up to 30 inches tall; its flowers are rose-purple. It has a seed pod that looks like a crane's bill before it pops open and scatters its seeds.

HABITAT: Wild geranium is found along roads and trails, in the forests, and in high meadows.

EDIBLE PARTS: The leaves can be boiled for a tea.

MEDICINAL USES: The leaves and the roots can be prepared in the following ways:

Leaves: Make a tea and use to soothe stomach troubles, give relief to tuberculosis sufferers, and wash sore eyes. Wild geranium contains tannic acid; therefore, the tea makes a good astringent.

Roots: Boil or soak roots in hot water. For sore throat, ulcers, or diarrhea, gargle with the tea or drink it. It also has been an aid to people with heart trouble. This tea is also given to new mothers and their babies as a tonic. The baby gets only a small amount.

For mouth sores, chew the raw root.

GOOSEFOOT FAMILY

LAMBSQUARTER, Pigweed, Goosefoot
Chenopodium

DESCRIPTION: One of the tastiest and best-known wild edibles, lambsquarter is an annual plant that grows from 1 to 4 feet tall. One of its common names, "goosefoot," comes from the fact that the leaves of many varieties resemble a goose's foot. However, lambsquarter growing in our area has a leaf that is long and pointed, fatter near the stalk, and with relatively smooth edges. These leaves are alternate and grayish-green underneath. Small, grayish-green flowers form in the joints between branch and stalk near the top of the plant, and in a cluster at the very top. These later turn into green-covered seeds.

HABITAT: Since pigweed, as it is called near Kodiak, or lambsquarter is an introduced plant, it is found where people have been. In our area it grows along many of the beaches, in the comparatively loose soil at the high tide mark. Look for it, too, in old garden areas and along roadsides.

EDIBLE PARTS AND NUTRITIONAL VALUE: The leaves, flower buds, flowers, and seeds are edible. This relative of beets and spinach is one of the tastiest wild greens. Lambsquarter contains high amounts of calcium, vitamins A and C, and protein, and significant amounts of thiamine, riboflavin, and niacin. It is more nutritious than spinach or cabbage.

MEDICINAL USES: The primary use of this plant medicinally would be as a dietary aid in diseases caused by calcium or vitamin A and vitamin C deficiencies.

WAYS TO PREPARE FOR EATING: Lambsquarter leaves can be used in salads or substituted in any spinach recipe, raw or cooked. They're especially good wilted with vinegar and bacon.

Steam the leaves for a short time; lambsquarter does not require a long cooking period, as leaves are tender and not at all bitter. They also make a flavorful creamed soup or purée.

Flowers and buds can be cooked until tender and served as a vegetable.

Seeds can be gathered in autumn, dried, and ground into flour. For baking, mix half-and-half with regular flour.

The whole seeds can be sprinkled over a salad or baked in bread, or boiled until they are soft and served as a cereal.

For additional recipes using lambsquarter, see Green Noodles from Mars and Summer Salad (both dandelion under Composite Family), and Wildwood Fritters (goosetongue under Plantain Family).

Pigweed Pie
Jim Tsacrios

3 eggs
½ pound feta cheese (more to taste)
⅓ ounce cream cheese
1 cup (packed) chopped, cooked pigweed (lambsquarter)
Freshly ground pepper
Dash salt
Pastry for double-crust, 12-inch deep-dish pie
Butter

Beat eggs, crumble feta cheese, cut cream cheese into small cubes; mix together. Stir in pigweed. Add pepper and salt. Pour into partially baked pie shell. Dot with butter; add top crust. Bake at 375 degrees until top crust is brown.

Variation: Nettles may replace pigweed.

. . . Lambsquarter
Tasty Foragers' Piroshkees
Ouzinkie Botanical Society

Pastry for 9-inch double-crust pie
3 cups cooked and chopped lambsquarter leaves
  (or any mild-flavored green)
1 carrot, finely chopped
¼ cup celery, finely chopped
¼ cup green pepper, finely chopped
1 egg, beaten
2 tablespoons mint leaves
1 teaspoon minced garlic
1 teaspoon salt

Roll pastry ⅛ inch thick. Cut in 3-inch squares. Mix other ingredients and place 2 tablespoons of this mixture on each pastry square. Fold dough over to form triangle; press the edges together. Place on cookie sheet and bake at 350 degrees until brown.

Summer Seed Dressing

1 cup mayonnaise
½ cup sour cream
½ cup choped petrúshki leaves and stems
3 tablespoons cut wild chives
1 tablespoon lemon juice
3 tablespoons tarragon vinegar
Dash pepper
¼ cup lambsquarter seeds

Blend mayonnaise and sour cream. Add petrúshki, chives, lemon juice, vinegar, and pepper. Add lambsquarter seeds while blending.

Wild Green Soufflé

1 quart leaves (lambsquarter or mixed greens)
2 eggs, beaten
1 cup thick white sauce
½ cup grated cheddar cheese

Cook leaves in 1 cup water until just tender. Leave in water until cool; drain. Add beaten eggs and white sauce. Put in casserole and cover with grated cheese. Bake at 325 degrees until cheese is browned — check after 15 minutes.

Forager's Fandango
Fran Kelso

1 pound ground round
1 medium onion, chopped
1 small can mushrooms
2 cloves garlic, crushed
1 teaspoon oregano
¼ cup salad oil
Salt
Pepper
2 cups (packed) steamed lambsquarter
   or boiled nettles, chopped
1 can cream of celery soup
1 cup sour cream (optional)
Mozzarella cheese, grated
¼ cup rice or noodles

Combine ground round, onion, mushrooms, garlic and oregano; brown in the oil. Add salt and pepper to taste. Stir chopped lambsquarter into meat mixture. Now mix in soup, sour cream and grated cheese to taste. Add rice or noodles. Put more cheese on top, sprinkle with paprika, and cook uncovered at 350 degrees for 30 to 40 minutes.

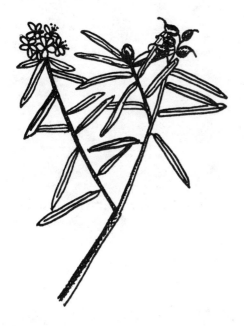

HEATH FAMILY

LABRADOR TEA, Hudson Bay Tea
Ledum palustre decumbens
Mogúlnik (Russian)

DESCRIPTION: This subspecies is a small relative of a shrub also called Labrador tea. It has long, narrow, needlelike leaves that stay olive-green on the upper surface the whole year. Leaf undersides are reddish brown. Little pink or white flowers form at the stem tops. The plant has a very pleasant aroma.

CAUTION: An effective laxative in large doses. At first, drink tea brewed from this plant in small amounts.

HABITAT: Labrador tea grows in woods; it especially likes bogs and swamps. Our class has found that it favors bog cranberries and crowberries as companion plants.

EDIBLE PARTS AND NUTRITIONAL VALUE: The leaves and branches, high in vitamin C, are brewed in "chai" (tea).

MEDICINAL USES: Prepare a tea by boiling fresh or dried leaves and branches until the water turns dark. Drink this tea for anemia, colds and tuberculosis. It can also be used for arthritis, dizziness, stomach problems, heartburn and hangover. This tea has been known locally as a remedy for chest ailments and tuberculosis.

WAYS TO PREPARE FOR EATING: This plant can be gathered year-round. It can be dried for winter, or dug up from under the snow.

Boil the leaves and branches until the water is the color of regular tea. Or add 1 tablespoon dried herb to 1 cup boiling water; steep 5 minutes and sweeten to taste. Makes a good hot drink; however, use sparingly to avoid laxative effect.

Native Alaskans made a meat spice and a marinade — for game with a strong wild taste — from this plant. The meat would be soaked in tea made from the boiled plant, or the meat, stems, and leaves would be boiled together.

HORSETAIL FAMILY

HORSETAIL, Scouring Rush, Joint Grass
Equisetum

DESCRIPTION: A perennial, horsetail is the only living member of a prehistoric family. Some of its ancestors grew to the size of trees. These mid-Paleozoic plants were in their prime 350 million years ago. (One reference notes that cockroaches have only been around 300 million years.)

The stalks of young horsetail are made up of a series of hollow joints. When young, the plant resembles an asparagus. Later on, the stalks develop many small branches and look something like horses' tails or miniature evergreen trees. By the time the small branches appear, the plants are no longer edible and should not be collected as a vegetable.

WARNING: If used as an internal medicine, take in small doses — a mouthful at a time, and no more than a cup a day. If taken in larger doses, or over a long period of time, horsetail can act as a poison.

HABITAT: Grows in damp woods and gravelly hillsides.

EDIBLE PARTS AND NUTRITIONAL VALUE: In the very early spring, just after the ground is thawed, dig underneath the stalks from the previous year. The new shoots, wrapped in a brown protective cover, resemble berries growing on top of the roots. Peeled, these are quite good to eat. In later

40 HORSETAIL FAMILY

. . . Horsetail
spring the upper stalk and head can be eaten. Do not eat brushy stalks.

This plant contains trace minerals.

MEDICINAL USES: The silica in horsetail makes it beneficial as a wash for sore eyes. For this purpose, squeeze out the juice from the part above the root (rhizome) and from the stem.

Or, make a tea and use it as a wash for wounds, sores, and skin problems, and as a gargle for mouth sores and inflamed gums. The plant can be dried and burned and the ashes put on mouth sores.

The acid from the silica in horsetail is said to stabilize the scar tissue in the lungs in cases of mild tuberculosis. Horsetail also promotes blood coagulation; thus it was a medicine for the internal bleeding of stomach ulcers. It also can help slow a heavy menstrual flow, assist in preventing water retention, and ease discomfort from urinary tract problems. It has been taken to treat and cure arthritis. To prepare horsetail, steep 2 teaspoons dried plant in ½ cup water.

Diedre Bailey of Kodiak gave us a recipe with horsetail that is beneficial to nursing mothers. This tea helps start the flow of milk. Again, it is the silica in the horsetail plant that makes it useful in this recipe. Mix together equal amounts of dried horsetail, chamomile, borage, and comfrey. Measure desired amount into a container: 1 heaping teaspoon of herbs per cup of tea. Add boiling water and steep 45 minutes. Add a dash of honey to sweeten, if desired. Diedre says the tea is tasty and effective.

OTHER USES: When the treelike part of the plant, called "scouring brush," appears, it makes an effective plate and pan scrubber for campers. This part of the plant also makes a greenish-yellow dye.

WAYS TO PREPARE FOR EATING: Peel the buds from the top of the roots and add to salads or stews. The upper stalk and head can be eaten raw or boiled. The head should be peeled and the inner core used. Do not eat brushy stalks.

For a recipe using horsetail, see Early Spring Salad (goosetongue under Plantain Family).

LILY FAMILY

INDIAN RICE, Chocolate Lily, Kamchatka Lily, Riceroot
Fritillaria camschatcensis
Laaqaq (Aleut)
Saraná (Russian)

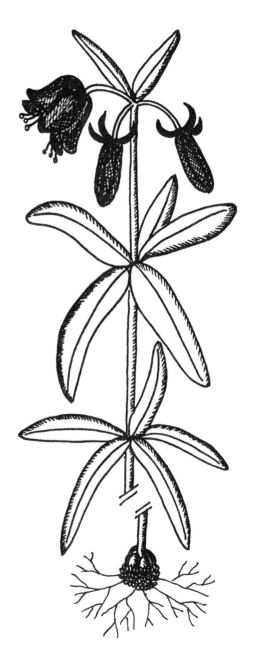

DESCRIPTION: Indian rice is a perennial that grows from a bulb made up of many ricelike bulblets. Its single stem reaches 1 to 2 feet tall. The leaves form circles around the stem, near the top, in 2 or 3 places. There may be 1 to 6 large, nodding, bell-like flowers. The flowers are dark purple to nearly black.

If you smell this flower you won't forget it, because it has a very disagreeable odor. Some of our class members have nicknamed it the "toe-jam flower."

HABITAT: Found in open coastal meadows.

EDIBLE PARTS AND NUTRITIONAL VALUE: The root can be eaten as a carbohydrate.

WAYS TO PREPARE FOR EATING: Dig Indian rice roots in the autumn. Break up the bulbs and soak bulblets in water overnight to help rid them of any bitter taste. They can be eaten raw with fish eggs.

To cook, boil the soaked bulblets like rice until they are soft (a little lemon juice added to the water helps remove bitterness). Try them in casseroles as a rice substitute, or mix them with oil and add them

. . . Indian rice
to soup or stew. The cooked bulblets are also good seasoned with tart berries such as raspberries.

For winter use, dry the bulblets and cook them in fish or meat stew later on, or pound them into a flour.

Indian Rice a la Russ
Russ Mohney
(from *Root, Stem and Leaf* by Glen Ray)

3 cups Indian rice
3 tablespoons butter
½ cup chopped onion
½ cup chopped celery
Mushrooms (optional)
1 cup leftover cooked pork, finely chopped
2 teaspoons soy sauce

Boil the Indian rice until it just begins to soften. Drain and set aside. In a large skillet, sauté the onion and celery in butter until transparent (the addition of a few sliced mushrooms at this stage doesn't hurt a bit). Stir in the cooked Indian rice and pork. Keep turning the mixture as you would fried rice until the Indian rice begins to brown in the butter. Add the soy sauce, stir well, cover, and lower the heat. Turn the Indian rice mixture occasionally until quite soft. Serves 6.

WILD CHIVES, Wild Onions
Allium schoenoprasum

DESCRIPTION: Smaller than the domestic onion, but looks, smells, and tastes like onion. The plant forms a rose-purple cluster, or head, of flowers in the latter part of the summer.

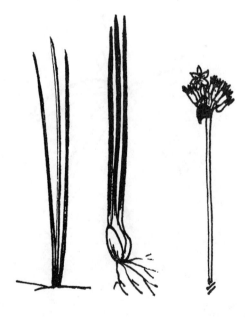

CAUTION: Don't eat any plant looking like wild onion unless it smells like onion. You might have an extremely poisonous plant by mistake (see death camas under Poisonous Plants).

HABITAT: Found in low meadows and pastures.

EDIBLE PARTS: Leaves and bulbs can be eaten.

WAYS TO PREPARE FOR EATING: Cut up leaves as seasoning — try in salads, stews, or casseroles. Start collecting leaves in early spring.

Bulbs can be harvested in late summer or early autumn. They have a strong flavor, so use only a small amount in place of domestic onions.

To store: Chop leaves. Then, in a container, put a layer of salt and a layer of greens, alternating until the container is full. Cover and store in a cool place. Or, chop the leaves fine with scissors, dry thoroughly, and store in spice bottles.

For recipes using wild chives, see Summer Seed Dressing (lambsquarter under Goosefoot Family) and Tabooley (petrúshki under Parsley Family).

WILD CUCUMBER, Watermelon Berry, Twisted Stalk
Streptopus amplexifolius
Oogóortsi (Russian)

DESCRIPTION: This is a perennial plant. The leaves are alternate, parallel-veined, longer than wide, and broad at the base. The small flowers are bell-shaped, white, pinkish or greenish; they hang from slender stems underneath the leaves. The berries are yellow-white to orange or red when ripe. The plant grows from 1 to 4 feet tall from thick, stringy, horizontal roots. A definite kink in each flower or berry stem near its middle gives the plant the name "twisted stalk."

LILY FAMILY

. . . Wild cucumber

HABITAT: Grows in moist woods and along stream banks.

EDIBLE PARTS AND NUTRITIONAL VALUE: The young shoots and leaves and the berries are edible. The succulent stem, leaves, and berries have a cucumberlike flavor.

Stems of wild cucumber sometimes grow to the thickness of young scallions. Gather these before they branch and add them to salads. They are juicy, with a nice texture — wild cucumber at its best.

These plants contain vitamins C and A.

MEDICINAL USES: Eaten in quantity, the berries act as a laxative.

WAYS TO PREPARE FOR EATING: The young shoots and leaves of wild cucumber are excellent in salads, creamed soups, and purées. The berries can be eaten alone or mixed with a little orange for jelly.

For additional recipes using wild cucumber, see Spruce Island Weed Salad (seabeach sandwort under Pink Family) and Spring Beauty Salad (Alaska spring beauty under Purslane Family).

Wild Cucumber Salad

4 cups (loosely packed) finely shredded wild cucumber leaves and chopped stems
1 large sweet onion
1 cup sugar
¼ cup mild vinegar
Salt
Pepper

Put greens in a bowl and cover with thin slices of onion. Pour the sugar over the top; allow to stand for a couple of hours in a cool place. Add vinegar, salt and pepper to taste, and serve.

MADDER FAMILY

NORTHERN BEDSTRAW, Alaskan Baby's Breath
Galium boreale

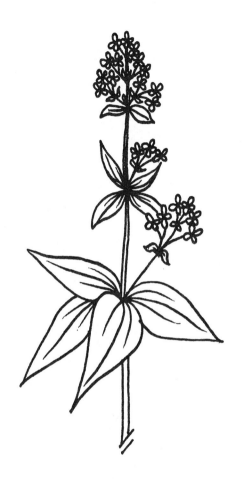

DESCRIPTION: This sweet-smelling plant grows about 3 feet tall. It has sprays of small white flowers and small, coarse leaves with a long, narrow shape. The leaves grow in groups of 4 under the flower sprays. The plant looks dainty and has a sweet fragrance.

HABITAT: Found in meadows, along roads and trails, and in gravelly areas.

EDIBLE PARTS: Young plants. Leaves, seeds and roots can be made into beverages.

MEDICINAL USES: Warm the plants in hot water and place externally on wounds. They help coagulate (clot) the blood.

Prepare as above and use as a hot pack for sore muscles. Brew a tea or squeeze out the juice of the fresh plant for a wash for skin problems, such as rashes, cuts, insect bites, infections, and blood poisoning. Apply the juice or tea daily as a wash, then destroy the cloth with which it was applied. Or make a salve for the skin by mixing the fresh juice with butter. Reapply this salve every 3 hours.

Bedstraw tea was taken internally for painful urination due to bladder infection or kidney stones. Steep 1 tablespoon dried bedstraw in ½ cup water. Drink 1 cup a day, made fresh each time.

OTHER USES: With an alum mordant, the leaves produce a yellow dye. A pink to purple dye can be made from the roots.

WAYS TO PREPARE FOR EATING: The young plants can be eaten as a cooked green — bring to a boil, drain, and boil again to eliminate any bitterness.

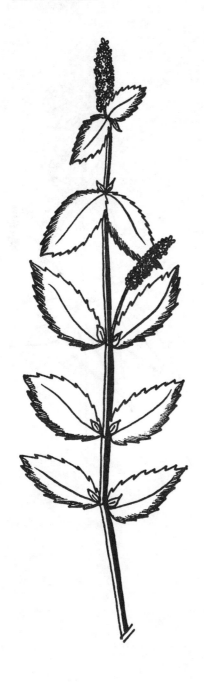

MINT FAMILY

SPEARMINT
Mentha spicata

DESCRIPTION: Spearmint is a rapidly spreading perennial which greatly resembles its peppermint relative. It has a square stem and opposite leaves, a bright green color, and gives off a definite minty smell. It is called spearmint because the leaves are pointed, like spear tips. Spearmint is not native to Alaska, although it grows well here when planted. This particular species was introduced due to its delightful aromatic qualities. It probably originated in Europe and was brought to North America by early European settlers. Spruce Island has its own largish patch of wild spearmint that jumped the fence from Sasha Smith's garden one year and has been flourishing in its new home ever since.

HABITAT: Mints like moist, rich soil.

EDIBLE PARTS AND NUTRITIONAL VALUE: The leaves are the edible part. They contain vitamin A.

MEDICINAL USES: Spearmint has the same properties as peppermint, although spearmint is a little less powerful. It can be steeped in tea to alleviate stomachache, gas pains, heartburn, and nausea, and to aid digestion. It can help calm nerves or prevent sleeplessness. It will also help ease menstrual cramps. Spearmint is excellent to give to children because it is mild.

To prepare, pour 1 pint of boiling water over about 1 ounce of the dried herb (make an infusion) and take 1½ to 2 cups a day.

OTHER USES: Spearmint and peppermint are often grown as commercial crops; the oil from these mints is used in medicines, kitchen products, cigarettes, cosmetics, toothpaste, bath items, candy and chewing gum.

WAYS TO PREPARE FOR EATING: Fresh or dried spearmint leaves can be enjoyed as tea (either alone or blended with other ingredients), in salads, sandwich spreads, with vegetables, for mint sauce and jelly, with soups and stews, and as a general seasoning.

Mint tea: Infuse about 1 ounce dried spearmint in 1 pint water. (When using fresh leaves, more than 1 ounce leaves is needed.) Or add 1 handful fresh mint to your "sun tea" for a delicious blend.

Salad: Mix 1 cup chopped leaves in a salad, tossing well; dress with oil and vinegar.

Open Sandwich: Mix chopped spearmint leaves with cream cheese and spread on whole-grain bread.

Vegetable: Stir 2 tablespoons minced mint into 1 quart peas. Or boil with new potatoes or cabbage.

Simple Mint Sauce: Combine 1 cup well-chopped spearmint leaves with just enough hot water to moisten them thoroughly. After the mass has cooled, stir in 1 cup orange marmalade.

Blender Mint Sauce: In a blender, combine the juice of ½ lemon, a scant ½ cup water, 2 tablespoons sugar, and 1 cup (packed) fresh spearmint leaves. Blend until smooth, chill for about 1 hour, and serve.

Soup and seasonings: Add powdered, dry leaves. Try it in potato salad or poultry dressing.

Spearmint Candy
Noreen Zeine

Dip fresh spearmint leaves in slightly whipped egg whites; dip in sugar; place on wax paper on cookie sheet. Dry in warm oven. Peel candy from wax paper and serve.

NETTLE FAMILY

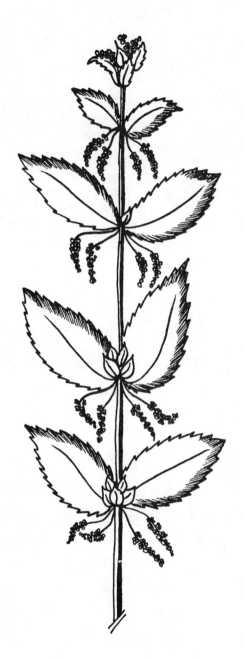

NETTLE FAMILY

NETTLE, Stinging Nettle
Urtica Lyallii
Uugaayanaq (Aleut)
Krapéva (Russian)

DESCRIPTION: A perennial plant, nettle grows from a single stem and varies in height from a few inches to 7 feet tall. Its leaves are dark green, opposite, coarsely grained, and sharply toothed. Stalks, leaflets and undersides of the leaves are fuzzy with fine stinging hairs (these contain formic acid). In late summer, long, slender, multi-branched clusters of green flowers bloom in the angles between the leaves and the stalks.

HABITAT: Found in deep, rich soil or near moisture; often in shaded places. Where nettles are found, sourdock or fiddlehead ferns are often growing as well. It is wise to pick nettles with gloves on. If you forgot to bring them, ease the nettle's sting by gathering sourdock leaves, squeezing out some juice, and rubbing it on the skin. Or rub the irritated area with the brown skin from fiddleheads.

Interestingly, around the Kodiak archipelago nettles are found growing *thick* around old barabara (early Native dwellings) and village sites.

My favorite nettle patch is on an open hillside, thick with ferns and false hellebore. The nettles seem to like the shade of these plants.

EDIBLE PARTS AND NUTRITIONAL VALUE: The greens are edible. The nettle is best early in the spring while it is still young and tender. Harvest the plant while it is less than 1 foot high.

Nettles are rich in protein, iron, vitamins C and A; they contain several minerals.

MEDICINAL USES: The nettle has many valuable properties. It acts as a blood coagulant, and, because of this quality, has been used to treat all kinds of internal hemorrhaging. It is helpful for diabetes, as it has been shown to lower the blood sugar level. A tonic prepared from the leaves can be taken for colds, headaches, and aftereffects of childbirth. Local sources say the tea was also given to aid those with tuberculosis.

For rheumatism, wash the place to be treated with hot water, then wrap with raw nettle leaves.

Ouzinkie people remember treating a toothache with nettle roots. Wash the roots, pound them, and hold them on the jaw in a heated rag as a poultice. Or bite down on the cleaned root, spitting out saliva.

An effective hair tonic can be made from nettles. For dandruff, simmer nettle greens in vinegar, cool, and massage into the scalp. For a hair rinse, make a tea of 5 handfuls of leaves to 1 quart boiling water. Allow this mixture to soak for several hours. Wash the hair and rinse thoroughly with the liquid produced by the nettle infusion. This rinse helps restore hair color and promotes hair growth.

OTHER USES: Rope can be made from the long fibers that run the length of the plant. The nettle has also been used to make paper, and a cloth that is said to be more durable than linen.

Seal hunters once rubbed themselves with nettles before going out to sea. The practice kept them awake during the long night.

Nettles will make a dye with color variations from yellow to bright green.

The plant is a commercial source of chlorophyll.

WAYS TO PREPARE FOR EATING: With scissors, cut the plants into pan-sized pieces. Boil them, then use like spinach. When cooked in boiling water, they lose their stinging quality in a very short time. Cooking time should be based on amount of

... Nettle
vegetables cooked and size and age of greens. Cook 5 to 15 minutes, just long enough to wilt thoroughly and make tender. Older plants are tougher and take longer to tenderize. If older plants are used, they can be boiled twice to overcome their stronger taste.

Nettle greens are good in fish soup, creamed soup, and purées. They can also be dried thoroughly, which eliminates their sting, and then crushed and measured for tea.

For additional recipes using nettles, see Pigweed Pie and Forager's Fandango (both lambsquarter under Goosefoot Family), Salmon and Nettles with Sunflower Seeds (Seafood chapter), and Deep-Dish Salmon-Wild Rice-Nettle Pie (Seafood chapter).

Alaskan Nettles
Sasha Smith

Gather nettles when plants are 2 to 4 inches tall. Wash nettles in several waters. Have saucepan ready. Pick nettles from rinse water; place directly in saucepan. The water clinging to the leaves will furnish sufficient cooking liquid. Cover; cook 5 to 10 minutes. Remove from heat; season with salt and pepper.

Variation: Cook 2 tablespoons chopped onion in 1 tablespoon oil; add washed nettles and cook until tender.

Nettle Casserole
Deborah McIntosh

Nettles
1 package dry onion soup mix
1 pint sour cream
1 cup shredded cheddar cheese
½ cup slivered almonds
1 tablespoon butter

Boil nettles 5 minutes or until tender, drain well, and chop into small pieces. Pack tightly to make 3 cups. Mix with soup mix, sour cream, and cheese. Sauté slivered almonds in butter until golden brown and sprinkle over the top of the casserole. Bake at 350 degrees for 30 minutes.

Nettle Loaf
Sandra Coen

1 pound nettles
Salt
2 eggs
½ cup milk
2 tablespoons butter
1 cup fresh whole-grain bread crumbs
¼ teaspoon pepper
¼ teaspoon salt
Hard-cooked eggs for garnish

Cut nettles in ½-inch pieces to make approximately 10 cups. Cook, covered, in as little boiling water as possible, salting to taste, for 5 to 15 minutes. Drain. Combine with remaining ingredients, except hard-cooked eggs, and turn into buttered loaf pan. Set pan in hot water and bake at 375 degrees for 25 minutes or until firm. Remove from pan and garnish with hard-cooked egg slices.

PARSLEY FAMILY

BEACH LOVAGE, Scotch Lovage
Ligusticum scoticum
Petrúshki (Russian)

DESCRIPTION: The triple-topped stems of petrúshki (as it is known in the Kodiak area) grow up to 3 feet tall. The leaves have long stalks; each leaf has 3 leaflets that are shiny and roughly toothed. The bottoms of the leafstalks and sometimes the very edges of the leaves have a reddish or purplish tint. The white or pinkish flowers of this perennial grow in flat-topped, umbrellalike clusters.

HABITAT: Found along sandy and gravelly seashores.

EDIBLE PARTS AND NUTRITIONAL VALUE: Leaves and reddish stems are high in vitamins A and C.

MEDICINAL USES: An infusion from this plant was prepared by people in the Kodiak area for kidney troubles.

WAYS TO PREPARE FOR EATING: Pick petrúshki in early summer before the flowers bloom. Ways of cooking with this plant are varied. The young leaves and stalks can be eaten raw in salads,

or substituted for parsley in any recipe. The plant can be boiled with fish or placed on baked fish. It can be used as a green vegetable; it makes a good cooked celery substitute. Petrúshki is added to soup stocks and stews, and used as a seasoning. The greens can be dried or frozen for winter meals.

Older plants might be bitter. Bring these to a boil, drain, then boil again.

For more recipes using petrúshki, see Boiled Wild Greens (sourdock under Buckwheat Family), Summer Seed Dressing (lambsquarter under Goosefoot Family), Elk Mulligan (Wild Game chapter), Deer Swiss Steaks (Wild Game chapter), and a number of recipes in the Seafood chapter.

Tabooley
Fran Kelso

1 cup bulgar wheat
¾ cup chopped onions
Salt
Pepper
1½ cups finely chopped petrúshki
¼ cup chopped wild chives
¼ cup chopped spearmint leaves
¾ cup olive oil
½ cup lemon juice
1 tomato, chopped
Iceberg or romaine lettuce leaves
1 tomato, cut in wedges

Soak bulgar for 1 hour in water to cover. Drain well and press out excess water. Mix bulgar with chopped onions and chives, add salt and pepper to taste. Crush onion mixture into bulgar with fingertips. Add petrúshki, spearmint leaves, olive oil, lemon juice, and chopped tomato. Serve on lettuce leaves and garnish with tomato wedges.

... *Beach lovage*
Deviled Eggs with Petrúshki
Georgia Smith

6 hard-boiled eggs
2 tablespoons mayonnaise
Dash vinegar
Salt
Pepper
Petrúshki
Sweet pickle
Paprika

Peel eggs and cut in half. Remove and mash yolks. Add mixture of mayonnaise, vinegar, salt and pepper. Chop petrúshki and sweet pickle in equal amounts, sufficient to double the volume of the yolk mixture. Refill the egg whites, sprinkle with paprika, and serve.

COW PARSNIP, Wild Celery
Heracleum lanatum
Ugyuun (Aleut)
Póochki (Russian)

DESCRIPTION: This perennial plant with grooved, very stout stems grows up to 9 feet tall. The leaves, large and divided into 3 leaflets, are shaped something like maple leaves. They grow from hairy stalks that are wider at the bottom and clasp the stem. The leaves have a white woolly look underneath and coarse, irregular edges. The white flowers are large, broad, umbrella-shaped, and flat-topped.

In the Kodiak area this plant is commonly known by its Russian name, póochki.

PARSLEY FAMILY   55

CAUTION: Do not confuse póochki with a similar plant, poison water hemlock (see Poisonous Plants chapter.)

Be sure to peel away the outer layer from the plant's stalk, as it contains a chemical that can cause blistering of the lips and skin irritation. The area of skin touched by the plant will be marked by large blotches, ranging in color from bluish to brown, sometimes swelling like hives. These blotches often remain for several weeks.

Although this chemical does not cause such reactions in everyone, it is best to be safe and gather

. . . Cow parsnip

póochki with gloves. Also, local sources say to gather the plant in morning or evening and not in the sunniest parts of the day. The chemical contained in the plant sensitizes the skin to light; thus this ingredient will not be as active when the sun is not as bright. Peeled, the inner stalk is safe to eat.

In the November, 1961 issue of *ALASKA SPORTSMAN®*, Eudora Preston contributed an article titled "Medicine Women." In the article, she gives a simple cure for the blotches or blisters caused by póochki. She says to bathe the affected skin with a mixture of vinegar and water, claiming it to be "the only sure cure I know."

HABITAT: Póochki grows in woods, fields, high meadows, and along the seashore. Though it is sometimes found growing in the shade, it prefers sunny places.

EDIBLE PARTS AND NUTRITIONAL VALUE: Inner stem and roots are reportedly high in sugar and contain some protein.

MEDICINAL USES: The root is the medicinal part. Here are some ways it has been prepared:

1) For colds, sore throat, mouth sores, and tuberculosis: Chew the raw root or boil it and drink the tea.
2) For arthritis, other body aches, swelling, cuts, and sores: Boil the root or soak it in hot water. Apply póochki water as a wash for the sore area, or mash the boiled root and place on the sore spot.
3) For toothache: Heat a piece of root until very hot and push it into the sore tooth. It will kill the pain by deadening the nerve.

OTHER USES: Póochki can also be used as a dye, according to Glen Ray in *Root, Stem and Leaf*. He says, "when used with alum, chrome, or copper mordants the flower heads yield a range of colors from light browns to yellow or gold."

WAYS TO PREPARE FOR EATING: When the plant is young, peel the outer layer of the stem and leaf stalks and eat the inner part the same way you would eat celery. The stem can also be cooked; some sources say cooking improves the flavor. A good way to serve it is as a replacement for celery in soups. Or, use it in creamed soups and purées.

Póochki leaves can be added to foods as a seasoning. Gather them in the autumn and dry them, then burn them by placing on a wire screen and holding the screen over a fire. Powder the ashes and store until ready to use.

The root was once eaten as a vegetable by Indians of the west coast of Alaska. It is said to taste something like rutabaga when cooked. A Larson Bay man told us his mother dug the roots each spring and pickled them in vinegar.

For an additional recipe using póochki, see Danny's Spicy Steamed Mussels (Seafood chapter).

Parsnip a la Hercules
(from *Root, Stem and Leaf* by Glen Ray)

6 cow parsnip (póochki) stalks, peeled
¼ cup boiling water
1 tablespoon sugar
Large dash salt
1 teaspoon lemon juice
2 tablespoons butter

Cut the peeled stalks into ¼-inch cubes. Put in a saucepan with all other ingredients except butter. Cook over high heat until all the water evaporates, keeping tightly covered all the while. Lower the heat, add the butter, and let the stalks brown. Serves 4.

Sweet Cow Parsnips
Russ Mohney
(from *Root, Stem and Leaf* by Glen Ray)

4 cups cow parsnip (póochki) roots
Salt
Paprika
¾ cup brown sugar or maple syrup
½ teaspoon grated lemon peel
1½ teaspoon lemon juice
2 tablespoons butter

Slice the roots into 1-inch pieces and boil until they are nearly tender. Place them in a shallow, greased baking dish. Season with salt and paprika to taste, then sprinkle with brown sugar. Top with lemon peel and juice and dot with butter. Bake, uncovered, for about ½ hour or until tender. Serves 8.

Variation: Mix a few slices of apple and some chunk pineapple with the roots before adding the brown sugar and lemon.

PEA FAMILY

BEACH PEAS
Lathyrus maritimus

DESCRIPTION: The beach pea grows in large clumps. Its leaves are thick and fleshy, with 6 to 12 oblong leaflets. The flowers, with a reddish banner and bluish violet wings and heel, resemble those of cultivated peas. The flowers are replaced by pods containing small peas.

CAUTION: Do not confuse with wild sweet pea *(Hedysarum Mackenzii)* or with lupine *(Lupinus nootkatensis)*. These plants are reportedly poisonous (see Poisonous Plants chapter).

HABITAT: Found along beaches.

EDIBLE PARTS: The peas are good to eat.

WAYS TO PREPARE FOR EATING: Eat the peas raw or cooked; add them to salads, or cook until just tender when pierced with a fork.

Try stuffing a salmon with a mixture of rice, onion, soy sauce, and beach peas, wrapping well with foil, and cooking on a grill over an open fire on the beach.

60 PEA FAMILY

CLOVER
Trifolium
Kléver (Russian)

DESCRIPTION: Most commonly these low perennials have three leaflets, growing from the plant stem on a slim stalk attached to the middle leaflet. The leaf edges are minutely toothed. The blossoms of white or red form dense balls.

HABITAT: Clover can grow anywhere that an open garden can be cultivated.

EDIBLE PARTS AND NUTRITIONAL VALUE: The flowers, leaves, and roots of all clover species are edible and high in protein.

MEDICINAL USES: For a good-tasting tonic, make a tea from clover blossoms which have been dried, broken into tiny pieces, and sealed in a jar to keep their fragrance. Add 1 heaping teaspoon of the crushed blossoms to 1 cup boiling water and steep 5 minutes. Add honey if desired.

Clover is good for curing skin diseases and for soothing sore throats and coughs. Clover contains tannin, which is an astringent.

Mayo Clinic research has shown that clover contains a blood thinner that may be useful for treatment of coronary thrombosis (heart disease).

OTHER USES: Clover roots contain bacteria that change nitrogen in the air into soil-improving organic compounds which help other crops grow. Clover is also a good food for grazing animals.

WAYS TO PREPARE FOR EATING: It is best to cook the plant, as it might cause gas if eaten raw. Steam the tender greens or use them as a potherb. The blossoms can be saved for brewing tea, making honey, or flavoring foods. The seed-filled dried blossoms can be baked in bread.

Clover root is also edible. It is best dug in the autumn or winter. Clean the root and remove the smaller fibrous parts. Chop into ½-inch chunks and boil 5 minutes. Drain, season, and eat as is or in stew.

For another clover recipe, see Fireweed Honey (fireweed under Evening Primrose Family).

Clover-Bright Salad
Russ Mohney
(from *Root, Stem and Leaf* by Glen Ray)

1 cup clover blossoms
2 cups dandelion leaves
12 mint leaves
1 onion, chopped
½ cup salmonberries or raspberries
1 medium cucumber, sliced
Mint sprigs
Oil and vinegar dressing or lemon

Soak clover blossoms overnight in salted water.
Tear dandelion leaves into salad-size pieces. Mix with mint leaves, onion, and dried clover blossoms. Place in a salad bowl and arrange berries and cucumber slices over the top. Put a sprig of mint on top for effect. Serve with a spicy oil and vinegar or other light dressing, or squeeze a fresh lemon over the salad and garnish with the peel. Serves 4.

PINK FAMILY

PINK FAMILY

CHICKWEED, Common Chickweed, Winterweed
Stellaria
Makrétzi (Russian)

DESCRIPTION: Chickweed is an annual or biennial which is found all over the world. It is usually a creeping plant. It has brittle stems, opposite oval leaves, and small, white flowers with 5 petals. The plant is yellowish green to green. Common chickweed is unique because it can start growing in autumn and later be found blossoming in winter. It got the name chickweed because it is a source of winter food for birds.

HABITAT: This plant grows in gardens, fields, lawns, waste places, and along roads.

EDIBLE PARTS AND NUTRITIONAL VALUE: The greens are the edible part. They contain some minerals and are also a good source of winter vitamin C.

MEDICINAL USES: For serious constipation, boil the herb (1 ounce chickweed to 1 pint water) and drink a large glass of the tea. Fresh leaves can be crushed and mixed with vaseline for use on bruises, irritations, and other skin problems.

WAYS TO PREPARE FOR EATING: Chickweed is tenderer than most wild greens, so eat it raw or cooked only slightly. It is a good green for salads when the plant is young. It is an excellent potherb, combining well with dandelions or watercress; it is also tasty wilted with vinegar and bacon. Chickweed is a good addition to soups and makes a flavorful creamed soup or purée. For a different breakfast treat, chop the greens and try them in pancakes some morning.

Fresh or dried, chickweed makes a refreshing tea.

Makrétzi Soup
Plants Class

1 small onion, diced
2 tablespoons (or more) butter
2 cups chopped chickweed
½ cup chopped sourdock
2 cups milk
1 potato, minced
Dash paprika
Salt
Pepper

Sauté onions in melted butter. Add chickweed and sourdock and cook until wilted. (Add more butter if pan gets too dry.) Mix in milk, minced potato, and seasonings to taste, and simmer until well-blended and thickened.

SEABEACH SANDWORT, Beach Greens, Sea Purslane, Sea-Chickweed
Honckenya peploides

DESCRIPTION: A bigger brother of chickweed, the seabeach sandwort is a perennial plant. It has smooth, sturdy stems with many branches that trail over the sandy beaches. It puts down roots at the stem joints as it spreads, forming dense mats of bright green along the shoreline. Its fleshy, succulent

. . . Seabeach sandwort
leaves are paired opposite each other. They are bright yellow-green and longer than they are broad. Small, five-petaled, greenish-white flowers grow either at the very ends of the leaf clusters or scattered among the upper leaves.

HABITAT: Grows on sandy beaches, starting just above the high tide point.

EDIBLE PARTS AND NUTRITIONAL VALUE: The leaves are the edible part. They are high in vitamins C and A.

MEDICINAL USES: Because of their ready availability to sea-going people and their high vitamin C content, the leaves have been eaten in the past by sailors as a cure for scurvy (a disease caused by lack of vitamin C). Arctic explorers gathered them to cure this disease among their crews.

WAYS TO PREPARE FOR EATING: Before the plant flowers, the young, juicy leaves can be eaten raw in salads; they have a sweetish taste. Or, mix them with other greens, such as mountain sorrel or sourdock, and make into kraut.

The Eskimos made a dessert by chopping the greens, cooking them in water, then allowing them to sour. The soured leaves were then mixed with reindeer fat and berries. The soured leaves were also eaten with dried fish.

Glen Ray, in his *Root, Stem and Leaf*, suggests cooking beach greens with sausage.

Franny's Favorite Spruce Island Weed Salad
Fran Kelso

Mix together equal portions of the following young plants, chopped in small pieces:

Seabeach sandwort leaves
Sourdock leaves
Fireweed
Wild cucumber

Add the following:

½ portion saxifrage leaves
½ portion spring beauty leaves
½ tomato, finely chopped
Minced onion (optional)
Minced garlic (optional)
Salt
Pepper
Crushed mint leaves
Blended salad herbs
Creamy dressing

Toss the salad. If desired, garnish with:

Salmonberry blossoms
Wild violet blossoms

PLANTAIN FAMILY

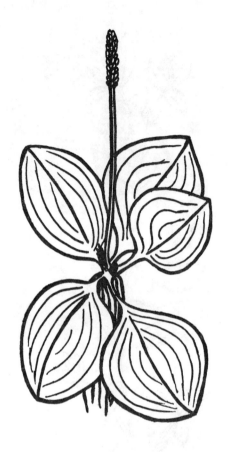

PLANTAIN FAMILY

BROAD-LEAVED PLANTAIN, Common Plantain, Snakeweed
Plantago major

DESCRIPTION: Plantain is a small perennial plant with broad, oval leaves. These leaves have several veins running lengthwise, from base to tip. Small white flowers grow from a brownish spike at the top of a central stem.

HABITAT: This plant is common in waste places, lawns, and along roadsides.

EDIBLE PARTS AND NUTRITIONAL VALUE: The edible leaves are rich in vitamin C and many minerals.

MEDICINAL USES: The whole plant is used as medicine. For a tea that is helpful for coughs, hoarseness, and general respiratory problems, or for gas pains, cover ½ handful of chopped leaves with 1 cup boiling water and steep ½ hour. Measure the same amounts and boil the herb in water for a decoction that helps coagulate (clot) blood.

The fresh juice can be squeezed from the leaves and taken for stomach problems or worms. The fresh leaves can be crushed and applied to cuts, sores, bites, and even hemorrhoids.

Chewing on the rootstalk can give temporary relief from toothache.

Sasha Smith tells us her father had bunions on his feet. Sometimes they would crack and become quite painful. When this occurred he would wrap his feet with plantain leaves, secure them with a bandage, and leave them on overnight. The plantain helped the bunions heal.

OTHER USES: According to Glen Ray in *Root, Stem and Leaf*, broad-leaved plantain leaves can be used for a dye. With alum, chrome, or copper mordants, colors vary from a greenish tint to yellow-brown.

WAYS TO PREPARE FOR EATING: Broad-leaved plantain and the narrow-leaved varieties (see goosetongue on the following page) can be prepared in the same ways, although the broad-leaved type is not as tender.

Leaves from any type of plantain are best when picked in the early part of the summer.

Eat the tender new leaves raw in salads. Or, cut the leaves into bite-size pieces and boil or steam them until just tender. Goosetongue leaves are also good wilted with vinegar and bacon.

Try to pick these plants young. If just older plants are available, cook the leaves with a cream sauce after puréeing and pressing through a seive.

Plantain Pizza

2 cups flour
1 teaspoon baking powder
2 tablespoons milk
¼ pound margarine
1 egg
1 small can tomato sauce
½ small can tomato paste
Salt
1½ quarts broad-leaved plantain or goosetongue, cooked (see "Ways to Prepare for Eating," following)
1½ cups grated Mozzarella cheese

Make a pizza dough of flour, baking powder, milk, margarine and egg. Spread on cookie sheet. Combine tomato sauce, paste, and salt. Spread over crust. Place alternate rows of plantain and cheese. Bake 30 minutes at 375 degrees.

PLANTAIN FAMILY

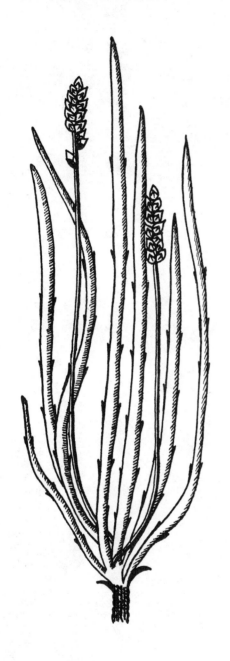

GOOSETONGUE, Seashore Plantain
Plantago macrocarpa, Plantago maritima

DESCRIPTION: Goosetongue, a perennial, has many long, narrow leaves rising from the base of the plant. The central stalk has a dense, blunt spike of flowers on the end. These flowers are very small, greenish or white, with 4 petals.

HABITAT: Goosetongue is found along the seashore, sometimes back some distance from the beach. The plant can also be found growing on rocky outcroppings near the water's edge.

EDIBLE PARTS AND NUTRITIONAL VALUE: The leaves are edible. They contain vitamins C and A, and some minerals. Goosetongue is a preferred spring food of Kodiak and Alaska Peninsula brown bears. They couldn't be wrong!

WAYS TO PREPARE FOR EATING: Goosetongue is prepared in the same ways as broad-leaved plantain — see page 67. Joyce Smith tells us her favorite method: Boil potatoes; during the last few minutes, place leaves on potatoes and steam until tender. Angeline Anderson says goosetongue makes a tasty addition to perok (salmon pie). For a tea, use ½ handful leaves to 1 cup boiling water; steep ½ hour.

Goosetongue Royale
Georgia Smith

4 cups (packed) goosetongue leaves
5 slices bacon, diced
3 tablespoons minced onion

Boil (or steam) goosetongue about 5 minutes. Drain. Fry bacon and onion in large skillet 3 to 4 minutes. Add goosetongue; sauté 2 to 3 minutes. Goosetongue Royale resembles green beans.

Wildwood Fritters
Ouzinkie Botanical Society

1 cup goosetongue leaves
2 cups lambsquarter leaves
1 cup sourdock leaves
2 eggs, beaten
2 tablespoons cottage cheese or yogurt
1 small onion, chopped fine
½ teaspoon blended salad herbs
Salt
Pepper
⅔ cup flour

Steam leaves until tender and chop into small pieces. Add eggs, cottage cheese or yogurt, onion, seasonings, and flour. Drop by spoonfuls on hot, oiled griddle and brown on both sides.

Early Spring Mixed Wild Salad
Jim Woodruff

1 cup goosetongue
1 cup sourdock
1 cup fireweed tips
1 cup dandelion leaves and blossoms
¼ cup horsetail
⅓ cup diced green onions
3 ounces bacon bits
½ teaspoon celery seed
½ teaspoon blended herbs
Salt
Pepper
Salad dressing

Wash and drain all greens thoroughly. Cut into bite-size pieces. Add green onions to blended herbs. Season with salt and pepper to taste; add dressing.

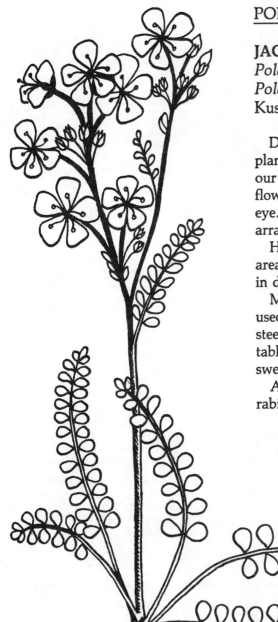

POLEMONIUM FAMILY

JACOB'S LADDER
Polemonium acutiflorum
Polemonium pulcherrimum
Kushelkok (Aleut)

DESCRIPTION: Jacob's ladder is a perennial plant. There are two varieties of Jacob's ladder in our area, one taller than the other. Both have blue flowers; the blossom of the short variety has a yellow eye. The plant got its name because of the ladderlike arrangement of its leaflets.

HABITAT: Jacob's ladder can be found in moist areas, but as it likes rocky terrain it is also found in drier soil.

MEDICINAL USES: The leaves can be dried and used as tea. Add 1 ounce to 1 pint boiling water, steep a few minutes, and drink as hot as is comfortable. This tea will help cleanse the body by causing sweating.

At one time Jacob's ladder was used to combat rabies.

PURSLANE FAMILY

ALASKA SPRING BEAUTY, Siberian Spring Beauty, Alaska Miner's Lettuce, Rain Flower
Claytonia sibirica
Lóstochki (Russian)

DESCRIPTION: Spring beauty is a small annual plant with a varying number of stems. Each stem bears a single pair of opposite leaves. Above the leaves sits a many-flowered cluster. The flowers are small with 5 white to rose-colored petals. In the Kodiak area, the blossoms are often pale lavendar.

According to Eric Hulten, *Claytonia acutifolia* and *C. tuberosa* do not grow in our area. These two members of the genus have edible roots which the Eskimos use like potatoes.

There is a local folk myth about spring beauty. People call the blossoms "rain flowers," and warn you that if you pick them it will rain. Someone around here must pick a *lot* of them!

HABITAT: Spring beauty grows along shores and on moist open hillsides. It likes wet places near running water.

EDIBLE PARTS AND NUTRITIONAL VALUE: Because the roots of spring beauty are slender, only the leaves are eaten. They are high in vitamins C and A.

WAYS TO PREPARE FOR EATING: Gather the sweet, tender young leaves in the spring. Add them raw to mixed salads or steam them for a short time and serve them as a green vegetable.

For another Alaska spring beauty recipe, see Spruce Island Weed Salad (seabeach sandwort under Pink Family).

Spring Beauty Salad

2 cups spring beauty leaves and stems
½ cup dandelion leaves
½ cup wild cucumber leaves
1 tomato, chopped
2 or 3 small green onions, chopped
Green pepper, chopped (optional)
Cucumber, chopped (optional)
1 clove garlic, minced
Salt
Pepper
Blended salad herbs
Dressing
12 to 15 violet blossoms

Chop spring beauty, dandelion, and wild cucumber into small pieces and combine with chopped vegetables and garlic. Season with salt, pepper, and blended salad herbs. Add oil and vinegar or creamy dressing and toss. Sprinkle with violet blossoms.

ROSE FAMILY

GEUM
Geum macrophyllum (Avens)

DESCRIPTION: Geum is a perennial herb with a stout, thick root. The first part of the plant to appear in spring is a rosette of many-leaved stems growing close to the ground. The leaves each have 3 lobes (curved or rounded parts). A large leaf grows at the end of the stem; a series of leaves, each smaller than the last, grows down the stem to the root. The leaves are fuzzy and the stems are hairy. As summer progresses, stems with several branches grow upward from the center of the plant. They have fewer leaves. Small, deep-yellow flowers appear at the stem ends. These are later replaced by round green seed balls that turn a brownish color with age.

The botanical name, *Geum*, comes from a Greek word meaning "to produce an agreeable scent." This refers to the aromatic roots. The dried rhizome (underground stem) of geum was once used as a clove substitute.

HABITAT: Grows in meadows, thickets, and wooded areas where the soil is moist and rich in nitrogen.

MEDICINAL USES: The dried rhizome, fresh flowering plant, or the leaves alone are used medicinally. Herbalists have brewed the plant into a bitter tonic used to increase the appetite after an illness or to help stop diarrhea. The tonic was also used as a gargle to soothe sore gums or to get rid of bad breath.

For a tonic, make a tea of boiling one ounce of the fresh plant in a pint of water. Drink the tea cold, a small glassful daily.

People in the villages in our area have used geum leaves as a pack for aches and sprains and pulled muscles. A local friend remembers using geum for such a purpose. She says to put the fresh leaves in hot water, then put them on the sore spot. They can be wrapped in a thin layer of cheesecloth first, if desired. Then wet a rag with very hot water and wring it out hard. Put it around the leaves. The moisture and heat will soak the beneficial parts out of the leaves and into the sore area.

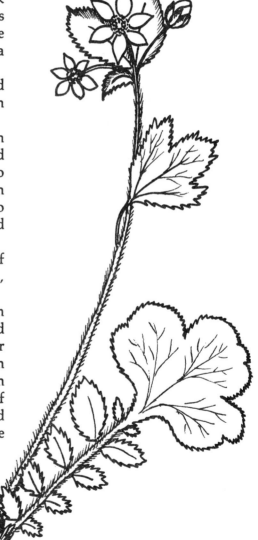

ROSE FAMILY

MARSH FIVEFINGER
Potentilla palustris

DESCRIPTION: This plant grows to 18 inches tall, often in a low, sprawling form. Its stems branch in "Vs". The leaves, palm-shaped with 4 to 5 long and slender leaflets, form near these stem junctions. Brownish-purple flowers form at the top of the stalk. Marsh fivefinger is the only member of the *Potentilla* genus that isn't yellow.

HABITAT: Grows along rivers and sloughs and at the edges of ponds. As its name implies, it likes wet, marshy places.

EDIBLE PART: The leaves are said to be dried for tea by the Eskimos. Plants Class has found them to be a satisfactory ingredient to use in tea blends.

SILVERWEED, Wild Sweet Potato, Silver Cinquefoil, Crampweed
Potentilla anserina

DESCRIPTION: Silverweed is a perennial with leaves that have saw-toothed edges. These leaves are bright green and smooth on top and silvery and woolly underneath. The plant produces strawberrylike red runners above the ground. It has single, large yellow flowers with 5 petals. The root is long and narrow.

HABITAT: This plant grows along seashores, lakesides and streams.

EDIBLE PARTS AND NUTRITIONAL VALUE: The edible root is a source of carbohydrates.

MEDICINAL USES: Make a tea of the fresh green leaves by steeping 1 teaspoon in 1 cup boiling water. Allow to cool and drink 1 to 2 cups a day to relieve diarrhea. Drink hot as a spring tonic.

To help gain relief from menstrual cramps, add 1 scant teaspoon silverweed leaves to 1 cup milk and scald. Drink warm. This remedy has also been used to relieve colic and aid digestion, to break a fever, and to help treat asthma and whooping cough.

To soothe toothache or sore gums, or to help tighten loose teeth, mix 1 ounce dried and crushed silverweed and 1 teaspoon alum with 1 pint white vinegar. Boil 5 minutes, or until the liquid is ½ to ¾ the original amount. Gargle with this blend for sore throat relief, or put it on freckles, skin blemishes, or sunburn.

OTHER USES: A reddish dye can be made from the roots.

WAYS TO PREPARE FOR EATING: Collect the roots in late autumn or early spring. Eat the roots raw or, preferably, boiled or roasted like potatoes. They taste a little like sweet potatoes.

Candied Silverweed Roots
Bradford Angier,
Feasting Free on Wild Edibles

1 pound silverweed roots, scrubbed
¼ pound butter or margarine
½ cup orange juice
½ teaspoon grated orange peel
¼ cup honey

Simmer roots in salted water until easily pierced with a fork. Slice the roots and layer in a baking dish with butter, orange juice and peel, honey, and salt and pepper to taste. Cover; bake at 350 degrees for 30 minutes, basting occasionally.

Silverweed Cakes
Bradford Angier,
Feasting Free on Wild Edibles

1 pound silverweed roots, scrubbed and boiled
Salt
Pepper
Butter
Bread crumbs

Mash the silverweed roots and add salt, pepper and butter. Form into cakes, roll in bread crumbs, and sauté in oil until browned.

SAXIFRAGE FAMILY

SAXIFRAGE, Salad Greens, Lettuce Saxifrage
Saxifraga punctata

DESCRIPTION: Saxifrage is a perennial. Its leaves are low-lying, with scalloped edges. The plant has a central flowering stalk that reaches from 4 to 20 inches tall. This hairy stalk has no leaves. The flowers at the stalk-tops are small, with 5 white petals.

HABITAT: Found in moist, rocky, shady places. The "saxifrage" portion of the name means "rock-breaker." Because it sometimes grows in cracks in rocks, it was supposed that it could split the stones. However, it doesn't create the cracks; it thrives in them because it is a small plant requiring little water.

WAYS TO PREPARE FOR EATING: Collect the leaves in the spring, before the flowers bloom. Saxifrage can be substituted in any lettuce recipe. Add leaves to salads, or fix as wilted saxifrage or saxifrage soup. Or cook with bacon and just a touch of sour cream.

For a recipe using saxifrage, see Spruce Island Weed Salad (seabeach sandwort under Pink Family)

STONECROP FAMILY

ROSEROOT, King's Crown, Scurvy Grass
Sedum rosea
Skrípka (Russian)

DESCRIPTION: This is a perennial plant named for its thick, fleshy root — it's rose-scented when bruised. Roseroot has numerous leafy, succulent stems, 4 to 12 inches tall. Its leaves are alternate, oblong or oval, with smooth or toothed margins. These fleshy, succulent leaves are pale green. The maroon, 4-petaled flowers form in dense heads at the ends of the stems.

HABITAT: Grows in rocky places and high meadows.

EDIBLE PARTS AND NUTRITIONAL VALUE: Leaves, stems and roots are edible.

This plant is a good source of vitamin C. Roseroot is one of several plants that have been called scurvy grass because of the vitamin C they provided to early travelers and settlers.

MEDICINAL USES: Brew the leaves or roots into a tea (1 ounce herb to 1 pint water) and drink to alleviate a cold, sore throat, or mouth sores. Or cool

VIOLET FAMILY

. . . Roseroot

the tea and wash a cut with it or apply it as an eyewash.

The fresh leaves have a cooling quality that soothes burns, bites, bruises, and other irritations. The leaves can also be used as a sponge for the eyes.

Chew the raw root and place it on cuts to aid in healing.

WAYS TO PREPARE FOR EATING: Harvest leaves and stems before plant flowers. They can be mixed, uncooked, in salads or cooked as a green. Where the plant is abundant, the roots can be boiled and eaten.

VIOLET FAMILY

WILD VIOLET
Viola
Fiálka (Russian)

DESCRIPTION: Violets are small perennial plants. Their heart-shaped leaves often stay green all year. Flowers can be white, yellow, bluish or purple.

HABITAT: Grow in moist woods or thickets.

EDIBLE PARTS AND NUTRITIONAL VALUE: The leaves and flowers can be eaten.

One-half cup of violets has as much vitamin C as 4 oranges. The plant is also high in vitamin A.

MEDICINAL USES: Violet tea is good for colds and sore throats. Heat the leaves briefly in boiling water and place on bruises to promote healing.

WAYS TO PREPARE FOR EATING: For a pretty, nutritious mixed salad, add violet leaves and flowers — or use them as a garnish. Both leaves and flowers can also be used in puddings and desserts. The leaves can be steamed and eaten.

Enjoy a tea made of leaves and flowers and sweetened with a little honey, or add a little of this mixture to a cup of regular tea.

For an additional recipe using wild violets, see Spring Beauty Salad (Alaska spring beauty under Purslane Family).

Violet Jelly
Sandra Coen

3 cups violet flowers
2 cups boiling water
Juice of 1 lemon
1 package liquid pectin
4 cups sugar

Put violets in mason jar with the boiling water. Let stand a few minutes, then strain. Add to violet water the lemon juice, liquid pectin, and sugar. Bring to rolling boil and boil 1 minute. Skim off foam and pour into hot, sterilized jelly glasses. Seal with paraffin and lids.

WATER LILY FAMILY

POND LILY, Yellow Pond Lily, Spatterdock
Nuphar polysepalum

DESCRIPTION: A perennial, the pond lily grows in water. It has a large, fat, spongy root that spreads along the lake bottom. The leaves, large and nearly heart-shaped, float on top of the water. Flowers are large and yellow, with touches of red; only one grows on each thick, spongy stalk. They are replaced by urn-shaped pods, an inch or two long. Inside each pod are several glossy, yellow or brown seeds.

WATER LILY FAMILY

... *Pond lily*

HABITAT: As their name implies, these lilies are found in ponds or at the edge of shallow lakes. Often the roots, a favorite food of beavers, are found on top of beaver lodges.

EDIBLE PARTS AND NUTRITIONAL VALUE: Roots and seeds can be eaten as a source of carbohydrates.

WAYS TO PREPARE FOR EATING: Gather the roots from autumn to early spring. Boil them twice or roast them, peel, and eat as a vegetable.

The seed pods are ready to be gathered in the autumn. Take the seeds from the pods, put on a cookie sheet, and roast in the oven at a low heat until they crack and the kernels can be removed. Then cook the kernels in the same way as popcorn. They will swell, and can be eaten as they are with salt and butter. They can also be ground into flour, or boiled and eaten as cereal.

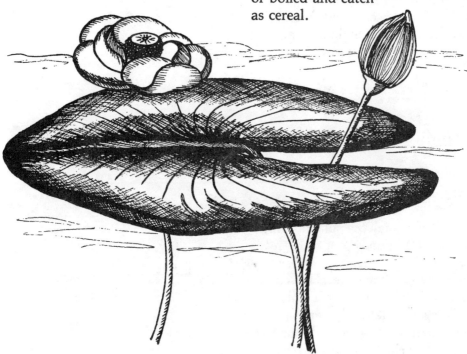

WINTERGREEN FAMILY

SHY MAIDEN, Star of Bethlehem, Single Delight, Wax Flower
Moneses uniflora
Kilitáyka (Russian)

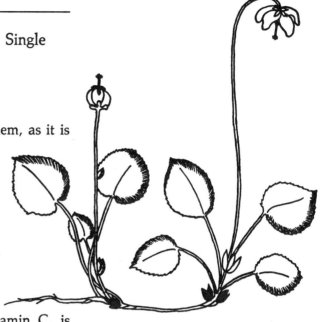

DESCRIPTION: The star of Bethlehem, as it is known locally, is a low plant with a cluster of leaves close to the ground. From this cluster grows a single stalk with a dainty white flower that faces down. The flower has a faint perfume.

HABITAT: Often found growing in the moss in the woods.

EDIBLE PARTS AND NUTRITIONAL VALUE: The whole plant, which contains vitamin C, is brewed for tea.

MEDICINAL USES: Prepare the whole plant, dried or fresh, as a general cold remedy. Add 2 or 3 plants to 1 cup boiling water and steep for 5 minutes. Try as a gargle for sore throat or drink for relief from other cold symptoms.

One local source says to bathe a sore or wound with tea from this plant, then cover the sore place with old, mildewy leaves dug from under salmonberry bushes.

In our area, this plant has also been dried and made into tea given for stomach disorders and lung troubles. It was known as a good medicine for treating tuberculosis.

Anna Opheim says she was taught that star of Bethlehem was good for rashes, bunions and corns. Pick the flowers, place them on the affected area, and tie them in place with a cloth bandage.

FERNS

BRACKEN FAMILY

BRACKEN FERN
Pteridium aquilinum
Paparótnik (Russian)

According to Eric Hulten, in *Flora of Alaska & Neighboring Territories*, the bracken fern is not found in the Kodiak area. However, it does grow in many areas of Alaska, and is included here for purposes of clarification, as it is the only fern that is considered dangerous to eat.

This fern has been eaten by people all over the world since ancient times. The young fronds are eaten while still coiled; they are known as "fiddleheads" because of their appearance. Indian tribes as well as early white settlers in Alaska cooked it as a vegetable. It has long been an important food in Japan, and is still sold as a commercial vegetable there.

However, research has demonstrated that bracken can be toxic. It contains a substance that stops the body from absorbing thiamine. If large quantities of this fern are eaten over a long period of time, death could result. Although it is unlikely that one would eat enough of the plant to cause poisoning, it is better to avoid this fern altogether.

There are many other species of ferns in Alaska, several of which can be eaten in the fiddlehead stage.

DEER FERN FAMILY

LICORICE FERN
Polypodium vulgare

DESCRIPTION AND HABITAT: The low-growing licorice fern usually is found on the trunks of alder trees. Its smooth fronds are less finely cut — with more space between "fingers" — than those of other ferns. Licorice fern stays green throughout the winter.

MEDICINAL USES: The stems of the licorice fern contain the same chemical that gives licorice candy its flavor. These ferns can be picked and chewed as a refresher for the tired hiker. They can be used raw or roasted as a cough medicine. The stems can be chopped, boiled, allowed to stand until cool, and taken for diarrhea or as a worm medicine.

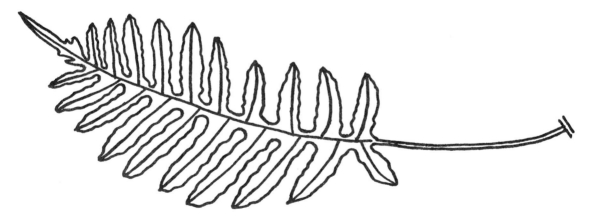

MAIDENHAIR FAMILY

MAIDENHAIR FERN
Adiantum pedatum

DESCRIPTION: The grayish-green fronds of this fern have a tufted appearance. *Adiantum* comes from a Greek word meaning "waterproof." The fern has this name because the leaves will not absorb much water. *Pedatum* means "having feet": the branching leaves of this fern look like birds' feet.

HABITAT: Found in the woods.

MEDICINAL USES: The leaves of the maidenhair fern can be chewed to help stop internal bleeding. A tea from the leaves is said to be good for treating coughs, hoarseness, and difficult breathing.

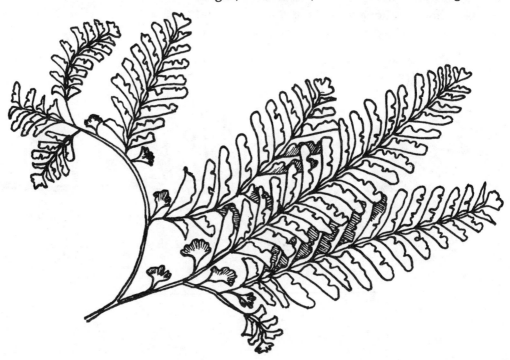

SHIELD FERN FAMILY

SPREADING WOOD FERN
Dryopteris dilatata

DESCRIPTION: This fern has an underground stem which is covered with the bases of the old leaf stalks. This underground part resembles a bunch of tiny bananas. When the fronds are full-grown they can be up to 1¼ feet long. Each frond is dense and shaped like a broad triangle. When the fronds first appear they are tightly curled and are called "fiddleheads."

HABITAT: They grow in moist woods.

EDIBLE PARTS: The bananalike part, which is underground, and the fiddlehead part, up to about 6 inches tall, can be eaten.

OTHER USES: The fibrous, dark brown roots are used for a brown dye. Fine roots are found lining nests of certain songbirds, such as thrushes.

WAYS TO PREPARE FOR EATING: Dig under the plant for the bananalike parts of the old stem, which grow right on top of the root. Roast them, remove the brown covering, dip the inner part in butter, and eat. This part of the fern is best in autumn, because it is juiciest then.

The fiddleheads are gathered in spring, before they get too high. As they unwind, they tend to become bitter. Remove the brown skin from very young fiddleheads and add them to salads. Try steaming fiddleheads and dipping them in a sauce of butter, lemon, salt, pepper, chili powder, and onion salt. Or, boil or steam fiddleheads and serve like asparagus, with butter or cream sauce. They're also good boiled 1 minute, drained, then sautéed.

The edible parts of wood ferns can be canned, or boiled, dried, and stored in a cool place until needed, then reconstituted with water.

. . . Spreading wood fern
Easy Fiddlehead Cheese Bake

4 cups fiddleheads
1 cup bread crumbs
2 eggs, beaten
1 cup grated cheese

Cook fiddleheads, draining and replacing water after it reaches a boil. When fiddleheads can be pierced easily with a fork, place in a casserole and mix in eggs, bread crumbs, and ½ cup of the grated cheese. Sprinkle remaining ½ cup cheese on top and bake at 325 degrees for 20 minutes.

BERRIES, LOW AND HIGH

To avoid confusion, we've supplied the following list of Kodiak-area berries, giving their names — common, Latin, Aleut and Russian — and showing their relationship to each other. You'll find more detailed information about each berry on the following pages.

Closely related berries, such as bog blueberry, black huckleberry, and early blueberry, can often be used interchangeably in recipes.

You'll find these versatile recipes for related berries positioned together, following the last description in the group of closely related berries.

Many of the recipes in this chapter call for sugar. In such a recipe, here or elsewhere in this book, honey in half the sugar amount can be substituted. Since honey will not cause jam or jelly from berries without natural pectin to thicken sufficiently, in these recipes add 1 level teaspoon agar per cup of liquid. (Agar is a kind of dried seaweed, available in most health food stores.)

CROWBERRY FAMILY

CROWBERRY, Blackberry
Empetrum nigrum
Shíksha, Shikshónik (Russian)

DOGWOOD FAMILY

BUNCHBERRY, Canadian Dwarf Cornel, Airberry, Dwarf Dogwood
Cornus canadensis

HEATH FAMILY

BLUEBERRY
Cuawak (Aleut)
Cherníka (Russian)

   1) **BOG BLUEBERRY**, Bilberry
   Vaccinium uliginosum

   2) **BLACK HUCKLEBERRY**
   Vaccinium uliginosum microphyllum

   3) **EARLY BLUEBERRY**, Forest Blueberry
   Vaccinium ovalifolium

CRANBERRY
Kenegtaq (Aleut)
Brusníka (Russian)

   1) **LINGONBERRY**, Lowbush Cranberry, Bog Cranberry
   Vaccinium vitis-idaea

   2) **LOWBUSH CRANBERRY**, Bog Cranberry, Swamp Cranberry
   Oxycoccus microcarpus

KINNIKINNIK, Mealberry
Arctostaphylos uva-ursi

HONEYSUCKLE FAMILY

ELDERBERRY, Red-berried Elder
Sambucus racemosa
Boozínik (Russian)

HIGHBUSH CRANBERRY
Viburnum edule
Amaryaq (Aleut)
Kalína (Russian)

LILY FAMILY

WILD CUCUMBER, Watermelon Berry
(included in Herbs chapter)

ROSE FAMILY

BEACH STRAWBERRY
Fragaria chiloensis
Zemlyaníka (Russian)

NAGOONBERRY, Wild Raspberry, Wineberry
Rubus arcticus
Puyurniq (Aleut)

RASPBERRY

1) **AMERICAN RED RASPBERRY**
   Rubus idaeus
   Malína (Russian)

2) **CLOUDBERRY**
   Rubus chamaemorus
   Maróshka (Russian)

3) **TRAILING RASPBERRY,**
   Mossberry
   Rubus pedatus
   Kostianíka (Russian)

SALMONBERRY
Rubus spectabilis
Alagnaq, Chughelenuk (Aleut)
Malína (Russian)

WILD ROSE
(included in Shrubs chapter)

CROWBERRY FAMILY

CROWBERRY, Blackberry
Empetrum nigrum
Shíksha - crowberry (Russian)
Shikshónik - crowberry bush (Russian)

DESCRIPTION: Crowberries grow on a low, trailing evergreen shrub with small, narrow leaves that look like spruce needles. The blossoms, small and purplish, grow singly or in clusters. The berries are juicy and black.

HABITAT: Crowberries grow in bogs.

EDIBLE PART: The berries can be eaten.

MEDICINAL USES: For relief from diarrhea, boil the leaves and stems of the crowberry and drink the tea. The cooked berries can be eaten for the same purpose. The berry juice, prepared as a drink, is said to relieve kidney troubles.

For sore eyes, a remedy can be made from crowberry roots. Clean roots thoroughly, boil them in water to make a tea, cool the liquid, and wash the eyes with it. Some Native people have been known to use the bark of crowberry stems to remove cataracts.

WAYS TO PREPARE FOR EATING: Crowberries are usually picked in the autumn, but are good all winter and into the next spring if they remain on the plant. By themselves they don't have much flavor, but are tasty mixed with other berries. They are a "watery" berry, lacking natural pectin, so mix especially well with blueberries. They are good in jelly and pies.

Crowberry Pie
Cooperative Extension Service,
Wild Berry Recipes

1 9-inch pie shell, baked
4 cups crowberries
1 cup sugar
1 tablespoon lemon juice
3 tablespoons cornstarch
¼ teaspoon salt
¼ cup water
1 tablespoon butter

Line the cooled pie shell with 2 cups of the berries. Cook the remaining berries with the sugar, lemon juice, cornstarch, salt and water until medium thick. Remove from heat, add butter, and cool. Pour over berries in the shell. Chill. Serve with whipped cream.

DOGWOOD FAMILY

BUNCHBERRY, Canadian Dwarf Cornel, Airberry, Dwarf Dogwood
Cornus canadensis

DESCRIPTION: Usually, many of these low-to-the-ground plants will be found growing together. The large leaves are oval with veins running their length, much like common plantain. These leaves are bright green and rather shiny. The flower grows from a single short stalk at the top of the plant just above the leaves. It has four white petals and a dark center. When the bright red, round berries form, they grow in tight clusters, several to a plant.

HABITAT: Bunchberries like spruce or birch forests, or alpine areas.

92 DOGWOOD FAMILY

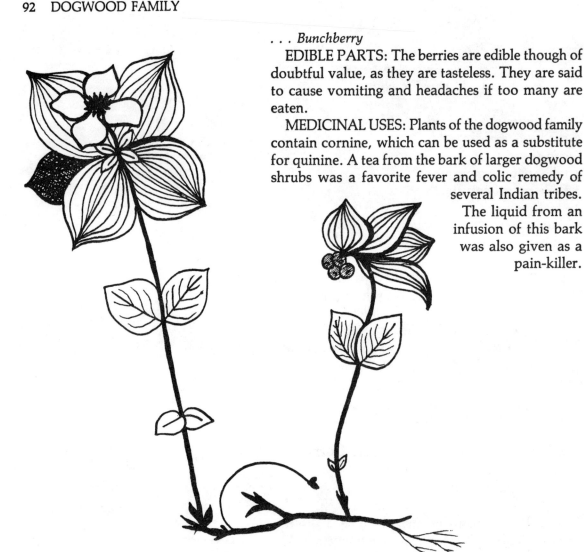

. . . Bunchberry

EDIBLE PARTS: The berries are edible though of doubtful value, as they are tasteless. They are said to cause vomiting and headaches if too many are eaten.

MEDICINAL USES: Plants of the dogwood family contain cornine, which can be used as a substitute for quinine. A tea from the bark of larger dogwood shrubs was a favorite fever and colic remedy of several Indian tribes.

The liquid from an infusion of this bark was also given as a pain-killer.

HEATH FAMILY

BLUEBERRY
Cuawak (Aleut)
Cherníka (Russian)

Of the seven closely related species or subspecies of blueberry growing in Alaska, three can be found in the Kodiak area. The berries can be used interchangeably in recipes. Blueberries produce a lavender to purple dye when boiled with alum, but the color tends to fade when exposed to sun.

1) BOG BLUEBERRY, Bilberry
Vaccinium uliginosum

DESCRIPTION: The bog blueberry, or bilberry, is a shrub with erect, branching stems. It is a small plant, growing 2 feet tall. Its leaves are small and alternate, rather thick, with smooth edges. It has tiny, bell-shaped, light pink flowers that grow from buds that form right on the old wood of last year's branches. The berry is blue-black.

HABITAT: This shrub can be found in bogs and marshy areas, and in the high country.

EDIBLE PARTS AND NUTRITIONAL VALUE: The berries are fairly high in vitamin C.

WAYS TO PREPARE FOR EATING: The berries can be eaten raw, cooked in various blueberry desserts, pies, puddings, pancakes, and muffins, frozen, or canned. These berries can also be used in jams, jellies or sauces. Recipes follow early blueberry, on the next page.

These fruits can be picked from early summer through the winter if you can find them then.

Bog blueberries in our area are sometimes wormy. However, if they are picked after the first heavy frost

. . . Bog blueberry
the worms will be gone. Or, if harvested earlier in the season, the berries can be soaked for 30 minutes in a saltwater solution. The worms will float to the top of the bowl and can then be removed.

2) BLACK HUCKLEBERRY
Vaccinium uliginosum microphyllum

This subspecies has generally the same description, habitat, and uses as the bog blueberry or bilberry (number 1 on preceding page).

3) EARLY BLUEBERRY, Forest Blueberry
Vaccinium ovalifolium

DESCRIPTION: This blueberry is a tall shrub, quite similar to other members of its genus. Its branches are stout and reddish-colored, and it has little, bell-shaped pink-white flowers. The fruit is round and blue.

HABITAT: It likes open woods, thickets, and slopes where there is peaty soil.

EDIBLE PARTS AND NUTRITIONAL VALUE: Berries are eaten and supply vitamin C.

WAYS TO PREPARE FOR EATING: Use in the same ways as other blueberries.

Blueberry Pie
Cooperative Extension Service,
Wild Berry Recipes

Pastry for 9-inch double-crust pie
3 cups blueberries
3 tablespoons flour or quick-cooking tapioca
1½ cups sugar
⅛ teaspoon salt
1 tablespoon butter

Mix all ingredients except butter and arrange in lower crust of pie. Dot with butter. Cover with the second crust and bake at 450 degrees for 10 minutes. Lower the temperature to 350 degrees and continue baking for 20 to 30 minutes or until juice bubbles up and the crust is browned.

Blueberry Pudding
Georgia Smith

3 cups fresh, fully ripe blueberries
1 cup sugar
¼ cup flour
1 egg, beaten
2 tablespoons butter

Place blueberries in medium saucepan. Mix sugar and flour with blueberries. Cook over low heat till juicy. Stir in egg and cook until thickened. Add butter and stir.

Blueberry Jelly

Blueberries
2 tablespoons lemon juice
7½ cups sugar
1 bottle liquid pectin

Crush fully ripe berries; place in jelly cloth and squeeze out juice. Measure 4 cups juice into *very large* saucepan. Add lemon juice and sugar and mix well. Place over high heat and bring to a boil, stirring constantly. At once, stir in liquid pectin. Then bring to a full rolling boil and boil hard 1 minute, stirring constantly. Remove from heat, skim off foam with metal spoon, and pour quickly into hot, sterilized glasses or jars. Seal with paraffin and lids. Makes 5½ pounds.

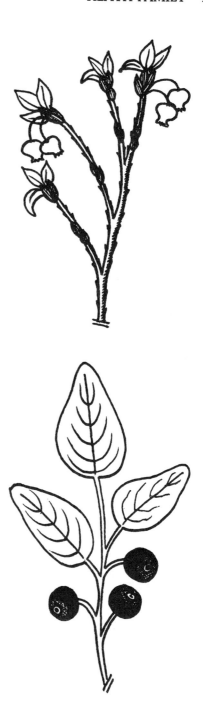

Blueberry Jam

4½ cups blueberries
7 cups sugar
2 tablespoons lemon juice
1 bottle liquid pectin

Crush fruit and measure 4 cups into a very large saucepan. Add sugar and lemon juice. Mix well. Place over high heat, bring to a full rolling boil, and boil hard 1 minute, stirring constantly. Remove from heat; at once stir in liquid pectin. Skim off foam with metal spoon. Stir and skim for 5 minutes to cool slightly (at high temperatures, fruit will float). Ladle into hot, sterilized glasses or jars and seal with paraffin and lids. Makes 6 pounds.

Wild Berry Fritters
Nancy and Walter Hall,
The Wild Palate

3 cups corn oil or safflower oil
4 cups rye flour
3 teaspoons baking powder
5 eggs
½ cup syrup made from berries
½ cup honey
3 cups any kind of berries (thawed if frozen, drained if canned)

Place oil in a deep kettle over medium heat. Heat gradually. Sift rye flour and baking powder together. In a separate bowl, combine eggs, berry syrup, and honey. Quickly stir egg mixture into dry ingredients. Don't beat. Gently fold in berries. When oil reaches 350 degrees, drop berry batter mixture in by tablespoons. Don't crowd. Turn fritters often until well done. Drain and serve hot.

CRANBERRY
Kenegtaq (Aleut)
Brusníka (Russian)

The two types of "cranberry" (*Vaccinium vitis-idaea* and *Oxycoccus microcarpus*) can be used interchangeably in recipes.

1) LINGONBERRY, Lowbush Cranberry, Bog Cranberry
Vaccinium vitis-idaea

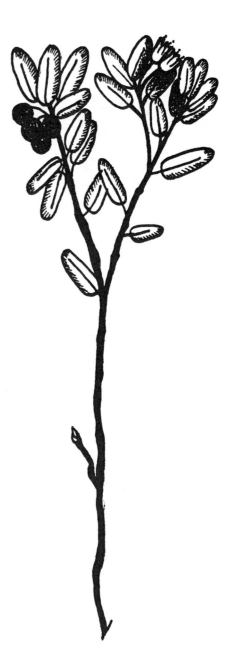

DESCRIPTION: This plant is a miniature, creeping evergreen shrub with slender, woody stems and tiny, shiny, dark green oval leaves with curled-under margins. The pink bell-shaped flowers grow either alone or in clusters and form at the ends of the stems. The berries are bright red and quite sour.

HABITAT: Grows in rocky places, in bogs, and on mountain slopes.

EDIBLE PARTS AND NUTRITIONAL VALUE: The berries are eaten; they contain vitamin C.

MEDICINAL USES: These cranberries, which grow all over Alaska, have been a Native remedy for headaches, swelling, and sore throats. They can be heated, wrapped in a cloth, and placed as a hot pack on the sore area, or raw berries can be chewed for sore throat.

OTHER USES: Boiling lingonberry leaves and stems with alum produces a red dye.

WAYS TO PREPARE FOR EATING: Preferably, pick after the first frost. These berries can be picked all winter if they're available. They have excellent flavor, and can be prepared in any way that commercial cranberries are used. They contain a considerable amount of benzoic acid, which will keep them from spoiling without sealing, whether

. . . Lingonberry

they are raw or cooked. Recipes follow lowbush cranberry, number 2.

These berries also can be made into an excellent cordial.

2) LOWBUSH CRANBERRY,
Bog Cranberry, Swamp Cranberry
Oxycoccus microcarpus

DESCRIPTION: These cranberries grow on a tiny evergreen vine with slender, creeping stems. The vine grows in the moss, putting down roots through its stem joints. The small, shiny green leaves are thick and leathery; the leaf edges roll under. Pink and yellow flowers at the top of erect stems look like miniature shooting stars. The berries are white at one end, shading from pink to a purplish red at the other end, where they attach to the stem. Some of our Plants Class members call them "pinkheads" because of their predominantly pink color.

Our references tell us that, in Alaska, these relatives of the commercial cranberry are not as plentiful as the lingonberry; in fact, they're downright scarce. However, on Spruce Island both varieties seem to be equally distributed. The small lowbush cranberries are very good and can be mixed with lingonberries.

HABITAT: Boggy or peaty soil is preferred by these small plants. They like to grow on top of sphagnum moss. I recently explored a newly discovered bog near my house and found hummocks of brown sphagnum moss so thick with "pinkheads" they seemed like tiny fields, planted there for me to harvest.

At first glance, it appeared that someone had thrown a large handful of these berries on the top of the sphagnum hump. When I looked closer, I noticed the berries were attached to a tiny vine which blended in with the moss underneath.

EDIBLE PARTS AND NUTRITIONAL VALUE: A source of vitamin C, the berries of this plant are eaten. Some Alaskans also eat the blossoms.

WAYS TO PREPARE FOR EATING: Can be used in the same ways as commercial cranberries.

Lingonberry Chiffon Pie
Cooperative Extension Service,
Wild Berry Recipes

1 cup lingonberry juice
1 package strawberry or lemon chiffon pie filling mix
⅓ cup sugar
1 9-inch pie shell, baked and cooled, or graham cracker crust

Chill ½ cup of the juice. Heat remaining ½ cup to boiling and add to pie filling mix in large mixing bowl; stir well. Add chilled juice and beat vigorously with rotary egg beater or electric mixer at highest speed until mixture stands in peaks — 1 to 3 minutes. Spoon into pie shell. Chill until set (about 2 hours). Serve plain or with whipped cream. *NOTE:* Store leftover pie in refrigerator, covered with inverted pie pan.

Cranberry Flip
Cooperative Extension Service,
Wild Berry Recipes

This creamy, pink, nourishing drink will delight everyone.

For each glass, allow 1 heaping teaspoon of vanilla ice cream and ⅔ cup cranberry juice. Beat with an egg beater or blender until light and frothy.

Cranberry-Banana Jam
Georgia Smith

3 cups cranberries
1½ cups water
2 cups mashed bananas
7 cups sugar
½ bottle liquid pectin
Lemon juice (optional)

Simmer cranberries and water for 10 minutes. Add bananas and sugar. Bring to a boil and boil 1 minute. Add liquid pectin. Stir and skim; add a little lemon juice, if you wish. Pour into hot, sterilized glasses or jars and seal with paraffin and lids.

Cranberry Jelly
Georgia Smith

4 cups fresh cranberries
1¾ to 2 cups water
2 cups sugar

Wash cranberries, removing stems. Place in 3½-quart saucepan. Add water; bring to boil. Reduce heat; simmer, covered, 20 minutes. Strain through cheesecloth. Bring cranberry purée to boiling point; boil, uncovered, 3 minutes. Add sugar; boil 3 minutes longer. Pour into hot, sterilized jars and seal with paraffin and lids. Makes 2¾ cups.

Cranberry Cake
Lea Bryan

2 cups flour
3 teaspoons baking powder
¼ teaspoon salt
1 cup milk
1 cup sugar
3 tablespoons butter
1½ to 2 cups cranberries (OK if still frozen)

Sift flour, baking powder, and salt together. Set aside. Cream butter and sugar; blend in milk. Add dry ingredients. Add cranberries. Spread batter in a square 8 or 9-inch buttered pan. Bake at 350 degrees for 40 minutes.

Topping:

1 cup sugar
1 tablespoon flour
½ cup butter
½ cup cream
1 tablespoon vinegar
1 teaspoon vanilla

In a saucepan, stir together sugar and flour. Add butter and cream and heat until bubbly. Add vinegar while beating. Add vanilla. Serve hot with warm cake.

Cranberry Muffins
Brenda Theyers-Wilson

⅔ cup unbleached white flour
⅔ cup soy flour
⅔ cup whole wheat flour
½ teaspoon salt
2 teaspoons baking powder
1 egg, beaten
2 tablespoons honey
1 cup milk
3 teaspoons melted butter
1 cup cranberries

Combine dry ingredients. In a separate bowl, mix egg, honey, milk, and melted butter. Blend with dry ingredients just until moistened. Stir in cranberries. Fill greased muffin tins two-thirds full and bake at 400 degrees for 20 minutes.

Crimson Cranberry Relish
Georgia Smith

2 oranges
4 cups cranberries
2 cups sugar

Wash oranges, cut into eighths, and remove seeds. Put oranges (including rinds) and cranberries through coarse blade of food chopper. Stir in sugar, blending thoroughly. Chill. Makes 6 cups.

KINNIKINNIK, Mealberry
Arctostaphylos uva-ursi

DESCRIPTION: Kinnikinnik is a creeping, low-growing shrub with reddish, woody branches that are only inches high. The shrub has leathery, oblong leaves tapered at one end. The bell-shaped flowers have 5 parts. They turn into a berrylike fruit that is dull orange-red.

Another species of this plant, *Arctostaphylos alpina*, or bearberry, also grows in our area but is not described here.

HABITAT: These plants like a well-drained, sandy soil. They can be found on open hillsides and in spruce forests.

EDIBLE PARTS: The berries are edible, though tasteless before cooking.

MEDICINAL USES: Leaves can be made into an astringent tea that is said to cleanse the kidneys. Pick young leaves and dry them at room temperature. Steep 1 teaspoon leaves in 1 cup boiling water for 5 minutes. Add honey if desired and drink warm.

OTHER USES: Dried leaves of kinnikinnik were the principal ingredient in a smoking mixture made by tribes of Northwestern American Indians.

WAYS TO PREPARE FOR EATING: These berries are tasteless raw, but become better with cooking. Mix with other berries and use the cooked juice for jelly.

HONEYSUCKLE FAMILY

ELDERBERRY, Red-berried Elder
Sambucus racemosa
Boozínik (Russian)

DESCRIPTION: The elderberry shrub can grow up to 14 feet tall. Its soft-barked stems grow straight. The leaves are opposite, with an uneven number of leaflets. The leaves have fine teeth along their edges and downy undersides. The bell-like flowers are pleasantly scented, small, and white. They grow in clusters at the ends of the branches.

WARNING: The seeds, leaves, twigs and roots of this plant are poisonous.

HABITAT: Found in woods and open areas.

EDIBLE PARTS AND NUTRITIONAL VALUE: Only the fleshy part of the berries should be eaten. Berries are rich in vitamin A, calcium, thiamine, and niacin, and are higher in calories and protein than other berries. Blossoms are also edible.

MEDICINAL USES: The Susitna people made a wash for infections by boiling the stem bark.

Although the root is said to be poisonous by some sources, Upper Inlet people boiled the inner root and drank the tea for colds, flu, high fever, and tuberculosis.

Local sources say that a tea made from elderberries is good for colds. The flowers, too, are good for this purpose. Dry the flower parts, boil them, and let the mixture cool. Drink this tea, a glassful twice daily, until the cold is gone.

A local lady says this tea is also just the thing when you get chilled and can't warm up. Drink the tea, stay inside, and cover up, as it will make you sweat.

If the berries or blooms aren't available, cut the stem and peel away the outer bark. The orange part inside can be made into a tea for the same purposes.

OTHER USES: Elderberry leaves make a yellow dye. The berries produce red, lilac blue, and plum to lavender dyes.

WAYS TO PREPARE FOR EATING: Boil the fleshy part of the berry for jelly. These berries are good mixed with a more acid-tasting fruit — try them with strawberries.

Elderberry Jelly
Cooperative Extension Service,
Wild Berry Recipes

4 cups elderberry juice
7½ cups sugar
½ bottle liquid pectin

Pour juice into a preserving kettle. Stir in the sugar. Place on high heat and, stirring constantly, bring quickly to a full rolling boil. Add the pectin and bring again to a full rolling boil and boil hard for 1 minute. Remove from heat. Skim off the foam quickly. Pour jelly immediately into hot, sterilized containers and seal with paraffin and lids.

Elderberry Wine
Fran Kelso

Pack a 1-quart measure with elderberry blossoms, pressing down firmly. Boil 3 gallons water with 9 pounds granulated sugar for 5 minutes, until a thin syrup forms. Add blossoms and mix well. Cool to lukewarm. Add 3 pounds chopped seedless raisins, ½ cup strained lemon juice, and 1 cake compressed yeast. Put into a large crock and let stand for 6 days, stirring 3 times daily. Strain and let stand for several months. Bottle or put into fruit jars. This light wine has the suggestion of a delicate champagne and keeps well for several years.

HIGHBUSH CRANBERRY
Viburnum edule
Amaryaq (Aleut)
Kalína (Russian)

DESCRIPTION: This tall shrub cousin of the elderberry, known as kalína in the Kodiak area, is easy to identify because of the noticeable musty odor of the plant — even jelly made from the berries has the scent.

Kalína branches are slender and gray, and can be 8 feet long. The leaves are opposite, shaped almost like maple leaves, with coarsely toothed edges. Small white flowers cluster on short branches.

The red or orange berries grow in juicy clusters that are easily gathered. Each berry contains a single flat stone.

HABITAT: This shrub grows from swamps to foothills. It likes meadows, open woods, and stream banks.

EDIBLE PARTS AND NUTRITIONAL VALUE: Kalína berries are edible and are very high in vitamin C.

MEDICINAL USES: The bark, pulp, and berries are used medicinally in the following ways:

Bark: Boil the inner bark and drink the tea for stomach trouble. In the Kodiak area and elsewhere in Alaska the inner bark of the branches has been brewed into a gargle for colds, sore throat, and laryngitis.

Pulp: After boiling berries to get juice for jelly-making, store the pulp in jars to use for colds, sore throat, or laryngitis. Then take 2 large tablespoons in 1 cup of very hot water to help break up the cold. Thelma Anderson treated me this way in Ouzinkie one winter night when I was ill. Not only is this Kalina concoction a healthy vitamin C drink — it's also a tasty one.

Berries: Eat uncooked for the same purposes. Raw or cooked, they have also been known as a remedy for tapeworms.

WAYS TO PREPARE FOR EATING: Though one of our class members claims "it smells like dirty socks," kalína jelly is still a tasty treat. Besides the berries, the only other ingredients needed for jelly are sugar and water, as the fruit contains its own thickening agent.

These berries remain on the plant well into winter and can be harvested whenever found. However, flavor is best if they are picked before the first heavy frost.

Highbush Cranberry Jelly
Georgia Smith

4 cups ripe highbush cranberries (kalína berries)
1 cup water
Sugar

Boil highbush cranberries and water for 3 to 5 minutes. When cool, strain through cheesecloth. Use ⅔ to ¾ cup sugar for each cup of juice. Boil approximately 10 minutes. Pour into hot, sterilized jars and seal with paraffin and lids.

Salmonberry-Highbush Punch
Rosemary Squartsoff

Makes a very good drink; tastes similar to tropical punch with a little tangy flavor.

To equal amounts of highbush cranberries (kalína berries) and sugar, mix in 4 to 5 times as much water. For 1 gallon of punch add 1 quart unsweetened salmonberry juice.

Highbush Cranberry Gelatin Dessert
Rosemary Squartsoff

To one 4-cup package of gelatin (peach, pear, and strawberry are best) add 2 cups boiling water. Stir well until dissolved. Add 2 cups highbush cranberry (kalína berry) pulp. Mix well; place in refrigerator to set. Makes a tangy dessert. *NOTE:* No sugar is added to berry pulp; any gelatin flavor can be used.

ROSE FAMILY

BEACH STRAWBERRY
Fragaria chiloensis
Zemlyaníka (Russian)

DESCRIPTION: A perennial plant, the beach strawberry has stout, thick, scaly roots. It starts new plants by runners, just like the cultivated strawberry. The leaves also look like those of its familiar garden cousin. They grow on long, slender stalks and have 3 leaflets with deeply toothed margins. The leaves are smooth on top and silky underneath. The flower, with its 5 white petals, blooms on a long, slender stalk. The fleshy, juicy red fruit grows up to one inch long.

HABITAT: Found on dry hillsides. We have heard it grows on the beaches on the Shelikof side of Kodiak Island. It was found on Spruce Island before the tidal wave in 1964.

ROSE FAMILY 109

EDIBLE PARTS AND NUTRITIONAL VALUE: Leaves, stems, stalks and berries are edible. Wild strawberries contain a great deal of vitamin C, iron, potassium, sulphur, calcium and sodium.

MEDICINAL USES: Can be eaten or used in tea to prevent vitamin C deficiency.

WAYS TO PREPARE FOR EATING: Traditionally, wild strawberries are eaten raw with sugar or used for jams, jellies, and other desserts.

For a tasty tea, gather over 2 handfuls of fresh green leaves, stems, and stalks, put into 1 quart of boiling water, and steep 5 minutes. Then serve plain or with fresh lemon juice and sugar. This tea is also good served cold the next day.

ROSE FAMILY

Wild Strawberry-Pineapple Conserve
Cooperative Extension Service,
Wild Berry Recipes

2 cups wild strawberries
2 cups canned crushed pineapple
2 cups sugar
1 cup pecans or walnuts

Mix strawberries, pineapple, and sugar; let stand 3 to 4 hours or overnight. Simmer slowly to develop the juice, then boil rapidly for 1 minute, stirring constantly. Remove from heat. Add nuts. Spoon into hot, sterilized jars and seal with paraffin and lids.

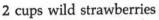

NAGOONBERRY, Wild Raspberry, Wineberry
Rubus arcticus
Puyurniq (Aleut)

DESCRIPTION: Locally known as wild raspberry, the nagoonberry is a low perennial plant. Its erect stems are less than 6 inches tall. Its leaves resemble those of the strawberry; they are divided

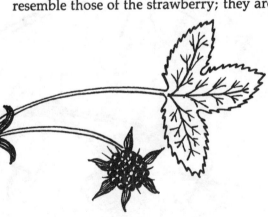

into 3 leaflets with coarsely toothed edges. The flowers, dark rose to red, are followed by berries made up of numerous small, juicy ovals. These berries look like raspberries.

HABITAT: Grows in damp, wet, relatively open woods or hillsides.

EDIBLE PARTS AND NUTRITIONAL VALUE: Young, peeled sprouts and, of course, the berries can be eaten. The fresh fruit, an extremely rich source of vitamin C, retains its high vitamin content if frozen immediately after picking.

WAYS TO PREPARE FOR EATING: Gather these berries in late summer. When found in sufficient quantity, the berries make an excellent jelly. They are also good in wines and cordials. A nonalcoholic beverage can be made by letting the ripe fruit stand in vinegar for 1 month, then straining out the juice. Dilute with water and ice, sweeten to taste, and serve on a hot day.

Nagoonberry Jelly
Cooperative Extension Service,
Wild Berry Recipes

5¾ cups nagoonberry juice
¼ cup lemon juice
6 cups sugar
½ bottle liquid pectin

Bring nagoonberry juice, lemon juice, and sugar to a boil (if some green berries are used, increase the sugar by 1 cup). Add liquid pectin. Bring to a full rolling boil and cook 1 minute. Pour into hot, sterilized jars and seal with paraffin and lids.

ROSE FAMILY

RASPBERRY

1) AMERICAN RED RASPBERRY
Rubus idaeus
Malína (Russian)

DESCRIPTION: A shrub with canes (stalks) 2 to 4 feet tall, this plant is an Alaskan favorite that has been transplanted to our area by immigrants from other parts of the state. Its branches are woody and brownish red. Its leaves have 3 to 5 roughly toothed leaflets that are whitish and hairy underneath. Its 5-petaled flowers form in clusters. The red fruit is made up of many balls formed into the traditional raspberry shape.

HABITAT: Grows in thickets, clearings, and along woods edges. Here on Spruce Island, various patches have been planted; these are assigned by a sort of seniority system to the local berry pickers, with no trespassing allowed. I did have a rare opportunity to help harvest an abandoned patch on Afognak Island last summer, and found the experience gratifying.

EDIBLE PARTS AND NUTRITIONAL VALUE: Berries are very high in vitamin C. They spoil easily and must be used or frozen quickly after picking. The leaves can be used for a tea.

MEDICINAL USES: Raspberry leaf tea has been used by pregnant women for hundreds of years to ease labor pains, prevent miscarriage, and increase their milk supply. Steep 1 ounce leaves in 1 pint boiling water for 15 minutes. Strain and drink at least 2 cups a day.

The tea can be taken to relieve diarrhea and used as a gargle for mouth sores and as a wash for cuts and sores. Eating the berries will also help stop diarrhea.

OTHER USES: Raspberry leaf tea may be beneficial as a hair rinse — it is said to be especially good for brunettes.

WAYS TO PREPARE FOR EATING: Use any raspberry recipe; see Clover-Bright Salad (clover under Pea Family).

2) CLOUDBERRY
Rubus chamaemorus
Maróshka (Russian)

DESCRIPTION: On this low perennial plant, the stems stand erect from a creeping root and grow to 8 inches tall. Two or three leaves grow from each stem; each leaf has 3 to 5 rounded lobes with toothed edges. Single white flowers with 5 white petals form at the ends of the stem. The pinkish-yellow, soft and seedy berry is made up of several small ovals clustered together like a raspberry.

HABITAT: Found in bogs and moist, relatively open areas.

EDIBLE PARTS AND NUTRITIONAL VALUE: The fresh berry is a very rich source of vitamin C. It should be cooked, eaten, or frozen right after picking.

WAYS TO PREPARE FOR EATING: Use cloudberries raw, in berry desserts, or in jelly.

3) TRAILING RASPBERRY, Mossberry
Rubus pedatus
Kostianíka (Russian)

DESCRIPTION: The trailing raspberry has a slender trailing stem. It roots at the stem joints. There are 5 toothed leaflets on each mature leaf. White-petaled, solitary flowers later form into 1 to 6 fruits shaped like small red balls. These berries are hard

ROSE FAMILY

. . . Trailing raspberry
to pick in quantity because they are small, grow in the moss, and are spread out over a large area.

HABITAT: Mossberries, as they are called locally, are found in woods and mossy areas.

EDIBLE PARTS: The berries are good to eat.

WAYS TO PREPARE FOR EATING: These small berries make a delicious jelly, if one has the patience to pick enough of them.

Mossberry Jam
Georgia Smith

4 cups mossberries (trailing raspberries)
2 cups sugar
Water

Combine berries and sugar in saucepan, adding enough water to keep mixture from sticking. Boil 10 to 20 minutes, depending on how ripe berries are. Pour into hot, sterilized jars and seal with lids. Process 15 minutes in a boiling water bath.

Wild Fruit or Berry Sauce
Nancy and Walter Hall,
The Wild Palate

2 cups whole berries (any kind)
1 cup wild fruit or berry juice
¼ cup honey
2½ tablespoons cornstarch
1 tablespoon lemon juice

Combine berries with juice, honey and cornstarch in saucepan. Place over medium heat. Stir in lemon juice, cook and stir 3 minutes. Chill thoroughly. Serve as dessert topping, over pancakes, or use in recipes calling for puréed fruit.

SALMONBERRY
Rubus spectabilis
Alagnaq (Aleut)
Chughelenuk (Aleut) — Young salmonberry shoots
Malína (Russian)

DESCRIPTION: Salmonberry is a many-branched shrub that can grow up to 7 feet tall. The stems are woody, with yellowish-brown bark. The bark peels off in thin layers. Leaves are made up of 3 leaflets. Deep pink flowers bloom from April to June. In June or July appear yellow to dark red berries.

There are three theories explaining how the salmonberry got its name:

1) The flowers are salmon-colored.
2) The berries look like salmon eggs.
3) The settlers used the bark to cure upset stomach brought about by eating too much salmon.

Salmonberry leaves contain chemicals that, when dissolved in water, kill the seeds of other plants. Therefore, few plants grow under the bushes.

HABITAT: These shrubs live on wooded mountainsides or in open, sunny areas.

EDIBLE PARTS AND NUTRITIONAL VALUE: Berries, blossoms, leaves and shoots are eaten or used in tea. The berries, especially, are an important source of vitamin C.

MEDICINAL USES: The bark and leaves have an astringent quality which is good for indigestion — steep them into a hot tea. Chewing on the young shoots can also aid digestion.

The undersides of green or dried salmonberry leaves can be placed on an infection to draw it out

ROSE FAMILY

. . . Salmonberry

or used on a wound that won't heal, say local sources. Some people dig out the old, mildewy leaves from under the bushes and place these over the sore or wound. This technique is used only with wounds that won't heal. Another method is to bathe the wound first in a tea made of Star of Bethlehem (Wintergreen family) and then apply the poultice of salmonberry leaves.

Leaves can also be chewed and placed on burns. Bark can be pounded and laid on an aching tooth or an infected wound to kill the pain.

WAYS TO PREPARE FOR EATING: Salmonberry blossoms make a pretty addition to salads.

The berries are eaten raw or cooked into jams, jellies, and berry desserts. They are also good in wines or cordials.

Young, tender shoots can be peeled and added to casseroles. They can also be coarsely chopped and sautéed in butter. These young stems were once a food source for the Tlingits and the people on the Kodiak archipelago.

Steep the dried leaves for a hot beverage: use 1 teaspoon dried leaves to 1 cup boiling water and steep 5 minutes.

For an additional recipe using salmonberries, see Clover-Bright Salad (clover under Pea Family).

Salmonberry Pie
Georgia Smith

1 cup sugar
About 3 tablespoons flour
3½ cups ripe salmonberries
Pastry for double-crust pie
1 tablespoon tapioca
Milk

Combine sugar and flour and add to salmonberries. Pour into bottom crust. Sprinkle with tapioca. Top with crust; brush with milk. Bake at 400 degrees for 30 minutes. NOTE: The juicier the berries, the more flour is required.

Salmonberry Jam

4 cups crushed salmonberries
7 cups sugar
½ bottle liquid pectin

Combine crushed berries and sugar in large saucepan. Place over high heat, bring to a full rolling boil, and boil hard 1 minute, stirring constantly. Remove from heat; at once stir in liquid pectin. Skim off foam with metal spoon. Stir and skim for 5 minutes, until mixture cools slightly (cooling prevents floating fruit). Ladle into hot, sterilized jars or glasses and seal with paraffin and lids.

SHRUBS

Shrubs which are known primarily as berry-producing plants are included in the preceeding chapter on berries. Two members of the Honeysuckle Family, elderberry and highbush cranberry, are described in that chapter.

GINSENG FAMILY

DEVIL'S CLUB
Echinopanax horridum
Cukilanarpak (Aleut)
Nizamýnik (Russian)

DESCRIPTION: The Latin name of this plant means "horrible weapon." Devil's club is a very prickly shrub with long branches, heavy with sharp spines. The "club" on the end of the stalk is totally spine-covered. The leaves are very large, shaped something like maple leaves, and also prickly. The berries are scarlet. Devil's club is a beautiful but potentially painful plant for the hiker, and a haven for the rabbit chased by the hiker's dog.

HABITAT: Found in the woods.

EDIBLE PARTS: For an emergency winter food supply in deep snows, dig to the roots and eat the new growth at the root tops.

MEDICINAL USES: Devil's club has a varied history of medicinal use by Native Alaskans.

Stems and branches: Cut into pieces and boiled into tea to treat fever.

Berries: The Haida Indians rubbed berries on their heads as treatment for lice and dandruff and to make their hair gleam.

Inner bark of the stem and root: Different Native Alaskan groups have found that the white pulp between the outer green bark and the stem acts as a laxative when chewed. The pulp was also used by Natives both internally and externally as a treatment for staph infections, and as a remedy for venereal disease.

The inner bark, either from the stem or roots, can also be applied to cuts. One method is to bake the root until very dry, then rub the pulp between the hands until it is broken up and quite soft. It may provide relief for swollen glands, boils, sores, and other infections. Leave the pulp on the area being treated for 3 to 4 hours only. It will burn if left on too long.

Roots: The British Columbia Indians prepared a strong tea from the root bark to treat diabetes. When clinically tested, it was found that an extract of devil's club lowered the blood sugar in rabbits.

The same tea, made by boiling the inner bark of the root in water, was used to treat people with tuberculosis, coughs, colds, stomach trouble, and fever. It is recommended that the tea be taken in small doses, as it seems to act as a strong stimulant. It is also suggested that this tea be made as a spring tonic. Tea made from roots gathered in autumn or winter may have toxic amounts of the active ingredients.

Devil's club roots, peeled of their stout outer layer, heated in the oven, and then mashed, can be placed on a sore area for relief of arthritis. The root and stem pulp can be put in bath water to ease the pains of rheumatism.

For a soothing pack for inflamed eyelids, burn devil's club roots and scrape off the resulting charcoal-like substance. Sieve it through gauze to make a fine powder. Moisten this powder with milk to make a poultice.

. . . Devil's club

It is said that when the bark is burned inside a house it purifies the air of disease.

OTHER USES: The oven-dried roots, ground fine, have been used as snuff.

ROSE FAMILY

MOUNTAIN ASH
Sorbus sitchensis

DESCRIPTION: Our type of mountain ash is a shrub that grows from 2 to 15 feet tall. It is an attractive and ornamental plant, with 7 to 11 leaflets springing directly from each leaf stalk. These leaflets have saw-toothed edges. The bark of the shrub is thin and grey. Many tiny flat clusters of flowers are replaced by clusters of vivid reddish-orange berries. These berries will stay on the mountain ash all winter if not picked. The leaves and berries have a strong smell.

HABITAT: The mountain ash likes moist or wet soil, and can be found on shaded slopes and in swampy areas.

EDIBLE PARTS AND NUTRITIONAL VALUE: Berries can be used as an emergency food supply. They contain vitamin C.

MEDICINAL USES: Because of their vitamin C content, the berries have been gathered to give to those with scurvy. These berries also contain an antibiotic, parasorbic acid, and have been effective as a gargle for sore throats and tonsillitis for some Alaskan Natives.

The Tlingits boiled the inner bark and took the tea as medicine. They considered it the very best thing for tuberculosis and severe colds. Kenai people soaked dried berries in hot water to make a tea for

the same purposes. Others have reported the bark, leaves, and dried berries helpful in treating ulcers, hemorrhoids and sores.

OTHER USES: Because of its astringent qualities, mountain ash has been used in tanning. The branches have been made into barrel hoops. Many people plant this shrub in gardens as an ornamental plant.

WAYS TO PREPARE FOR EATING: Some Indians ground dried berries into meal and flour in years when food was scarce. The berries can also be made into jams, jellies and marmaladas, though preferably mixed with other fruits. Wine can also be fermented from the berries.

WILD ROSE, Prickly Rose
Rosa nutkana
Róza (Russian) — Rose
Shipóynik (Russian) — Wild Rose Bush

DESCRIPTION: Wild rose is a shrub with many thorns on the stems. The canes, or shrub branches, grow from 1 to 4 feet tall. The branches have alternate leaves made up of 3 to 9 narrow, oval leaflets. These leaflets are smooth on top and downy underneath, with toothed edges. The flowers grow alone or in clusters of a few, usually have 5 petals, and are rose-pink. The "hips" are behind the flowers. They look something like little apples when they are ripe.

HABITAT: Wild rose grows in thickets and on rocky slopes or open hillsides.

EDIBLE PARTS AND NUTRITIONAL VALUE: Petals, leaves, and hips can be eaten. The hips are so high in vitamin C that foods made from them will retain enough vitamin content for winter use. Three

. . . Wild rose

little rose hips contain as much vitamin C as 1 orange. The rose hip seeds contain vitamin E.

MEDICINAL USES: The petals can be soaked in hot water and used to wash sore eyes. Boil a dark tea from broken-up stems and branches and drink to alleviate colds, fever, and stomach troubles, and to start menstrual flow. Peel the bark and soak in hot water until tea is very strong; drink this liquid if it is necessary to cause vomiting.

People here have used both petals and rose hips to help cure colds, especially when there is a persistent cough. Brew petals and rose hips into tea and continue taking regularly until the cough goes away.

OTHER USES: Rose petals can be put in talcum powder, potpourris, rose oil and perfume.

WAYS TO PREPARE FOR EATING: Gather leaves and petals in summer when roses are in full bloom, dry on a screen, and store for use in tea. The leaves are boiled in water for tea — 1 teaspoon of dried, crushed plant per 1 cup of boiling water — and the petals are added to 1 cup of regular tea or used in aromatic blends.

Fresh petals can be added to salads. And, fresh rose petals can be kept frozen for several weeks before using in jelly.

Rose hips are gathered in the autumn or winter when they are red. They should be prepared soon after they are collected. Wash them, remove the "tails" (the leafy part attached to the end), cut them in half, partially cover with water, bring quickly to a boil, and then simmer for about 15 minutes. Strain through cheesecloth to remove the juice. Store this juice in a cool place until it is made into syrup or jelly.

Sieve the pulp to remove seeds and skins and put in jams, marmalades, and ketchups.

Grind the seeds, boil, and use the juice in combination with juice from the hips to utilize the vitamin E contained in these seeds. Or use this fluid in place of water in jelly recipes.

Rose hip juice is good combined with that of a tart fruit (for example, bog cranberries) when making jelly. The juice can also be fermented to make wine.

Rose Hip Jam
Reprinted by permission © *The Herb Quarterly*, Newfane, Vermont 05345

Rose hips
Tart apples
Handful of rose petals (optional)
1 lemon, thinly sliced
½ cup water
5 cups sugar

Gather rose hips before they grow soft and cut off the heads and stems. Slice them in half and discard seeds and pithy flesh. Cover with water, cook until soft, and press through a sieve. Measure the purée. To each 4 cups purée add 1 cup peeled and finely chopped tart apple and, if desired, a handful of rose petals. Cook sliced lemon in ½ cup water for 15 minutes. Drain the liquid into the rose hip purée. Add sugar, bring mixture to a boil, and cook slowly until thick. Pour into hot, sterilized jars and seal with paraffin and lids.

Rose Hip Syrup

4 cups rose hips
2 cups water
2 cups sugar

Remove stems and flower remnants from rose hips; wash thoroughly. Boil rose hips and water for

. . . Wild rose

20 minutes in a covered saucepan. Strain through a jelly bag to clear the sediment from the mixture. Return the clear juice to the kettle. Add sugar to the juice and boil the mixture for 5 minutes. Store in a refrigerator or jar until you use it. Keeps indefinitely.

Rose Hip Lemonade

2 cups water
1 cup mild honey
3 teaspoons rose hips
¾ cup lemon juice

Heat water in saucepan. Stir in honey, add rose hips, and steep 1 hour. Strain and stir in lemon juice. Store in glass container in refrigerator.

To make 1 glass, mix ¼ cup base with ¾ cup water. Makes 12 cups.

Rose Petal Jelly
Georgia Smith

1 cup (packed) rose petals
1 cup water or apple juice
2 tablespoons lemon juice
Sugar
2 drops red food coloring (optional)
½ bottle liquid pectin

Simmer rose petals, water, and lemon juice until petals have lost their color. Strain the liquid to remove petals. Measure the strained liquid and add ¾ as much sugar as there is liquid. To tint the juice a rose color, put in red food coloring. Add liquid pectin. Boil rapidly for 1 minute, then pour into hot, sterilized jars and seal with paraffin and lids. Store in cool, dark place.

Rose-Chicken Waldorf Salad
Reprinted by permission © *The Herb Quarterly*,
Newfane, Vermont 05345

2 cups diced apples
2 tablespoons lemon juice
2 cups diced cooked chicken
1½ cups finely chopped celery
1 cup chopped walnuts
½ cup white raisins
½ cup sour cream
3-ounce package cream cheese, softened
1 cup (packed) rose petals
Lettuce

Toss apples with lemon juice. Combine with chicken, celery, walnuts and raisins. Combine sour cream, softened cream cheese, and rose petals in blender; pour over chicken mixture and toss until coated. Chill in refrigerator and serve on lettuce leaves.

Rose-topped Cheesecake
Reprinted by permission © *The Herb Quarterly*,
Newfane, Vermont 05345

Crust:
1⅓ cups graham cracker crumbs (about 16 squares)
3 tablespoons sugar
3 tablespoons butter or margarine, softened

Filling:
3 (8-ounce) packages cream cheese, softened
1 cup sugar
1 egg, warmed to room temperature
1 teaspoon vanilla

ROSE FAMILY

. . . Wild rose
Topping:
1 pint sour cream
3 tablespoons confectioner's sugar
1 teaspoon vanilla
1 cup red or pink rose petals

Preheat oven to 350 degrees. In a small bowl, mix together graham cracker crumbs, sugar, and softened butter. Press crust mixture evenly over bottom and sides of 9-inch springform pan. Set aside. In a large bowl, combine cream cheese, sugar, egg, and vanilla and beat thoroughly. Pour filling carefully into crust and bake for 30 minutes, or until cracks show on the surface. Cool. Combine topping ingredients in a blender and spread over cheesecake. Chill thoroughly before serving.

WAX MYRTLE FAMILY

SWEET GALE, Bog Myrtle
Myrica gale L.

DESCRIPTION: Sweet gale is a shrub with leaves that fall off in winter. Its reddish branches grow almost vertically to 3 feet tall. The gray-green, oblong leaves have a pleasant spicy smell. Sweet gale has brown and yellowish-green catkins (the flowering part — something like pussy willow catkins). These catkins are replaced by many small, flattened berries.

HABITAT: Sweet gale is commonly found in bogs.

EDIBLE PARTS: Leaves and berries can be dried and used in small amounts as seasoning.

MEDICINAL USES: The leaves were boiled into a tea and given by some Native Alaskans to aid tuberculosis sufferers. The tea was also known to be good as a wash for boils and pimples. At one time the tea was applied externally to treat scabies.

OTHER USES: A branch of sweet gale can serve as a steambath switch.

The dried, crushed leaves can be scattered in an infested area to repel and destroy insects such as fleas. A safe, nonchemical flea collar can be made by sewing a liberal quantity of the crushed leaves into a soft cloth strip and putting this collar on your pet.

The leaves can also be crushed for a sachet, or scattered in the bottom of bureau drawers to give contents a pleasant smell.

Sweet gale berries contain myrtle or myrica wax, which is similar to that contained in the berries of its wax myrtle or bayberry relatives. This wax can be added as scent to homemade candles.

Dye can be made from this plant. The roots and stem bark dye wool yellow; the leaves make a golden yellow, warm yellow brown, or cool yellow-green dye.

WAYS TO PREPARE FOR EATING: A small amount of dried leaves or berries can be used as seasoning in meat dishes.

Sweet gale was popular at one time in northern Europe as a flavoring for beer.

TREES

BIRCH FAMILY

ALDER
Alnus crispa
Wainiik (Aleut) — Parts of bush or tree used in banya (steam bath)
Veyníki (Russian)

DESCRIPTION: The alder is usually not a big tree. Its branches may be stout and long, but often they turn and twist every way but straight up. Its bark is grayish with dark spots.

HABITAT: Alder can be found in a variety of locations. It likes to take over an open hillside.

EDIBLE PART: The cambium, or inner layer of the bark, can be eaten.

MEDICINAL USES: The inner bark has been used for several purposes. Boil it and drink the tea to reduce high fever or to get rid of gas. This tea may also be taken as a gargle for a sore throat or laryngitis. This same inner bark, boiled in vinegar, may be effective as a wash to kill lice or for other skin problems. Dried alder bark acts as an astringent.

The female flower clusters, which don't look like flowers at all, are green, then brown late in the year; oblong; and less than 1 inch in length. They can be boiled into a tea and taken in small quantities for relief from diarrhea.

In the plant-screening program of the National Cancer Institute it was found that red alder, which does not grow in our area of Alaska, contains two anti-cancer agents. We were unable to determine if our species of alder was also tested.

OTHER USES: Alder wood is good for smoking fish or as a hot-burning firewood. When smoking fish, peel outer bark from alder before burning the wood — this bark can cause an unpleasant aftertaste in the smoked fish.

Alder branches are cut for steambath switches.

The bark of alders can be made into a brown dye; with mordants, ice green, orange, and gray-brown dye colors are possible. In the areas of Alaska where red alders grow, a reddish dye can be made from the bark.

Alder is the first tree to grow back on a logged or burned area; it was the first tree to grow in Alaska after the glaciers retreated. It serves an important purpose in future plant development because it adds nitrogen to the soil.

PINE FAMILY

SPRUCE
Picea sitchensis
Shíshki (Russian) — Cones

DESCRIPTION: The spruce tree is our most common tree having needles for leaves. It is an evergreen. There are heavy stands of Sitka spruce in the Kodiak area; Spruce Island got its name from these forests.

HABITAT: Grows in all forests in the Kodiak area.

EDIBLE PARTS AND NUTRITIONAL VALUE: The cambium, or inner layer of the bark, can be eaten. Indians in British Columbia used to mix this inner pulp with cranberries to form cakes which they ate either fresh or dried.

The spruce tree contains vitamin C.

... Spruce

MEDICINAL USES: There is a long history of use of the spruce tree for medicinal purposes, among both the Native people of our area and those in other parts of the state. Listed below are some of the uses:

Sap: Has been prepared as medicine for tuberculosis. Peel off a small section of bark in the spring, scrape off the sap, and chew it. This sap can also be placed on a burn or cut to help heal it.

Cambium, or white inner layer of bark: Cut off the outer, coarse bark, cut strips of the white inner bark, and boil or soak in hot water. The tea made from the strips is good for colds, sore throat, or mouth sores. It has also been a remedy for heart problems, kidney disorders, ulcers, and other stomach illnesses. It can be used as a wash for new babies. Don't drink this tea in large amounts unless you need a laxative.

These white strips of the inner bark can also serve as temporary bandages to stop bleeding from cuts.

Needles: Boil these for a tea that will cleanse your system. In smaller doses this tea is an effective cough medicine.

From the new growth of the needles, a spruce beer can be made that is good for cold sufferers or for tuberculosis patients. The needles are boiled and strained and beer brewed from the resulting liquid. This beer was a popular alcoholic beverage with the early sourdoughs. They collected the soft green tips of spruce boughs in the spring and fermented the brew from them.

The uncooked juice from these new needles can be squeezed into sore eyes.

Cones: When the new cones of the spruce are almost prime, they are light green and soft. Boil these and drink the tea for chest colds.

OTHER USES: A brown dye can be produced by boiling the item to be dyed with peeled spruce bark. Boiling with chopped twigs yields a camel-hair tan shade. "A long time ago, before there were kickers (outboards)," says a local man, fish nets were dyed this way.

The shíshki, or spruce cones, are fine fire-starters. The soft pitch can serve as a caulk for boats, while the hardened pitch makes a chewing gum. Baskets and hats can be formed from the roots.

Spruce wood has the highest strength-to-weight ratio of any wood. Therefore, it is in demand by many carpenters.

An excellent paper can be made from spruce because its long fibers create a strong newsprint.

WILLOW FAMILY

COTTONWOOD
Populus balsamifera
Ciquq (Aleut)

DESCRIPTION: The cottonwood is a big, tall tree which has thick, rough bark. In the summer the seeds float through the air in a fluffy, white, cottonlike material. These seeds gave the cottonwood its name.

HABITAT: Cottonwoods like to grow in open meadow areas and moist soils.

EDIBLE PARTS: The sweet inner layer of the bark can be eaten in the early spring.

MEDICINAL USES: The crushed leaves can serve as an antiseptic. For a sore throat, boil the bark and gargle the liquid. The bark can also be ground into a drying powder for sores.

OTHER USES: Cottonwood is excellent to burn for smoking fish. Because it likes wet places, it is a good indicator that water is nearby.

In other areas of the country, cottonwood is being farmed for paper production.

WILLOW
Salix
Vérba (Russian)

DESCRIPTION: Several willow species grow in Alaska. The catkins or "pussy willows" from a willow tree are a familiar sight, and may help in identifying the willow to the early spring forager. Willows can be shrubs or small trees. The leaves vary in size depending on the species; however, they are generally long and narrow. There are often many branches on this plant.

HABITAT: The willow likes moist ground. Look for it along streams and rivers.

EDIBLE PARTS AND NUTRITIONAL VALUE: All willows have edible parts. The leaves, buds, new sprouts, and inner layer of the bark can be eaten.

Willow is 7 to 10 times richer in vitamin C than the same quantity of orange.

MEDICINAL USES: To combat fever or gain relief from a stomachache or headache, peel off the bark from new willow growth and chew the inner layer. Willow bark contains salicin, which is a close relative to the pain reliever found in aspirin.

Bruised willow leaves, applied to cuts and wounds, help promote healing. They also have an astringent effect (they help reduce discharge or secretions from body tissue).

OTHER USES: Boil willow leaves and young twigs in water for a hair rinse to remove dandruff. The bark can be made into string. Baskets made of willow bark will hold water without caulking because the wood swells when it is wet.

WAYS TO PREPARE FOR EATING: Young willow leaves can be eaten raw with an oil dressing. For a vitamin C boost, suck the juice from young stems or munch on young leaves or new sprouts that have been peeled of their outer layer.

The Eskimos scraped the inner bark and ate it with sugar and seal oil. It has a somewhat sweet taste so doesn't need much sugar. This same inner bark can be cooked in strips like spaghetti or dried and pounded into flour.

POISONOUS PLANTS

Avoid the plants in this chapter; they are poisonous. (As are some of the plants elsewhere in this book, unless used with care.) Use the plant descriptions in this chapter to help you distinguish dangerous plants from similar but safe ones.

CROWFOOT FAMILY

BANEBERRY
Actaea rubra

DESCRIPTION: Baneberry is a perennial with a thick root and stems that are smooth and somewhat hairy. These stems grow from 2 to 3½ feet tall. The leaves are large; their edges have several divisions and coarse teeth. Small white flowers cluster in a spike at the top of the stem. There are 4 to 10 small, white petals on each flower. The berry is rounded and usually shiny red, rarely white. Inside are many seeds. These berries, about the size of a pea, are attached to the stem by a short, thick stalk. They are strikingly attractive to little children because they resemble small cherries.

HABITAT: Baneberry grows in the woods and along the edges of moist streams.

POISONOUS PARTS AND CONDITIONS OF POISONING: Both the roots and the berries contain poison. It can cause increased pulse rate, dizziness, vomiting, bloody diarrhea, gas pains, burning in the stomach, and trouble with breathing.

CROWFOOT FAMILY

BUTTERCUP
Ranunculus

DESCRIPTION: A perennial that grows from ½ to 2 feet tall, the buttercup is the first yellow-flowering plant in the spring. The plant often has fine hairs. Leaves sprout at the plant's base and (sometimes sparsely) along the stem. The leaves have deep indentations or are made up of more than one part. They are often very dark green. The yellow flowers at the end of the stalk have fine, *shiny* petals.

HABITAT: Buttercup grows in moist or wet soils in a variety of regions. The Latin name comes from a word meaning "frog," probably because many of these plants grow in standing water.

POISONOUS PARTS AND CONDITIONS OF POISONING: Many of the *Ranunculus* species contain a poison which will cause severe digestive problems. It is said that in some species the poison is destroyed by cooking, but it is safer not to try eating them.

MONKSHOOD
Aconitum delphinifolium

DESCRIPTION: Monkshood is a perennial. It has a straight, thin stem that grows to 40 inches tall. Its few leaves, separated into 5 toothed "fingers," look a little like delphinium leaves. A few dark blue flowers grow at the top of the stalk. The flower blossoms have a rounded

. . . Monkshood
hood. Occasionally one can find monkshood with white blossoms.

HABITAT: Monkshood grows in meadows, thickets, and along creeks.

POISONOUS PARTS AND CONDITIONS OF POISONING: The whole plant, including roots, is *extremely poisonous*. Monkshood contains a substance that paralyzes the nerves and lowers the body temperature and blood pressure.

Eskimos used to put poison obtained from this plant on the tips of their spears for killing whales.

NARCISSUS-FLOWERED ANEMONE,
My Darlings
Anemone narcissiflora

DESCRIPTION: This plant is a perennial with silky, hairy stems which grow up to 2 feet tall. The leaves, attached with long stalks, sprout mostly at the base of the stem. A few leaves grow higher on the stem, just below the flower clusters. The silky, hairy leaves are 1 to 5 inches long with many divisions. Flower clusters have white petals, often with a blue tint on the back.

HABITAT: Found in open meadows and on hillsides.

POISONOUS PARTS AND CONDITIONS OF POISONING: It is said that people in the Aleutian Islands sometimes eat the early spring growth at the top of the root of this plant. However, some members of this plant family contain a substance that causes sickness in animals who feed on it.

IRIS FAMILY

IRIS, Wild Flag
Iris setosa

DESCRIPTION: The iris, or wild flag, is a perennial that grows to 30 inches tall. The plant has long, wide-bladed, grasslike leaves. Large flowers at the top of the stalks are blue or purple, shading to white centers. Sometimes white blossoms are found, though these are rare. There are 3 big drooping petals on each flower.

These flowers were considered weather omens in local folklore. Sasha Smith remembers being told when she was little that it would rain if she picked wild iris flowers.

HABITAT: Iris grows in meadows and bogs and along streams.

POISONOUS PARTS AND CONDITIONS OF POISONING: The whole plant is poisonous and causes vomiting if eaten. There are various known medicinal uses, but these are dangerous to try unless you are an expert.

LILY FAMILY

LILY FAMILY

DEATH CAMAS
Zygadenus elegans

DESCRIPTION: This lily family member is a perennial growing from an onionlike bulb. Its long, grasslike leaves clasp the stem. The stem itself is smooth and grows from 1 to 2 feet tall. The greenish-white flowers grow in a long cluster at the top of the stem. Each flower has 3 petals.

HABITAT: Death camas likes open, dry areas such as meadows, roadsides, and forest edges. This plant does not grow in the Kodiak area, but is included for the benefit of other Alaskans who may be using this book.

POISONOUS PARTS AND CONDITIONS OF POISONING: All parts of the plant contain a poison similar to that contained in false hellebore (following). The poison affects the nervous system. Symptoms are salivating and nausea, vomiting, lowered body temperature, abdominal pain and diarrhea, difficulty in breathing, and coma.

When not in bloom, death camas might be mistaken for wild onion. Smell it first — if it doesn't smell like an onion, don't pick it!

FALSE HELLEBORE, Corn Lily
Veratrum viride

DESCRIPTION: A perennial plant, false hellebore has a stout stem, 3 to 8 feet tall, growing from a thick root. Leaves are alternate, 6 to 15 inches long, and broad. They are roundly oval in shape, with a pointed tip, and they enclose the stem. These leaves are folded lengthwise like the pleats in a skirt. The flowers are small and greenish with 3 petals; they gather in large, spikelike clusters at the top of the stem.

HABITAT: False hellebore grows in bogs, meadows, and creek bottoms.

POISONOUS PARTS AND CONDITIONS OF POISONING: The whole plant is poisonous. It can cause salivating, vomiting, diarrhea and abdominal pain, weakness, general paralysis, and spasms which sometimes become convulsions. The alkaloids in the plant can slow heart rate and lower blood pressure. In fact, the plant has been used medicinally for these purposes, and some other purposes as well. However, it is best to avoid this plant because it can be dangerous if used incorrectly.

Some people have become poisoned by false hellebore, by mistaking it for skunk cabbage — an edible wild plant which false hellebore resembles when it is young. Skunk cabbage, by the way, does not grow in the Ouzinkie vicinity, although it has been found on Kodiak Island near the village of Karluk.

PARSLEY FAMILY

POISON WATER HEMLOCK
Cicuta Douglasii
Cicuta mackenzieana

DESCRIPTION: Poison water hemlock is a perennial with a stem 3½ to 7 feet tall. This stem is stout, jointed, and hollow between the joints. *Cicuta mackenzieana* leaves are alternate, divided into narrow leaflets up to 4 inches long. The edges are toothed; the leaf veins end near the tooth notches. The leaf stalks sheath the stem. The leaves of

PARSLEY FAMILY

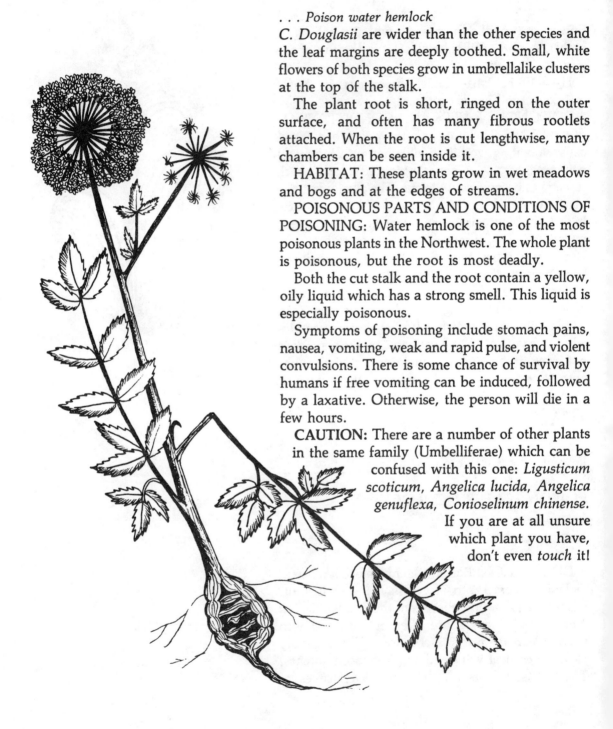

... *Poison water hemlock*

C. Douglasii are wider than the other species and the leaf margins are deeply toothed. Small, white flowers of both species grow in umbrellalike clusters at the top of the stalk.

The plant root is short, ringed on the outer surface, and often has many fibrous rootlets attached. When the root is cut lengthwise, many chambers can be seen inside it.

HABITAT: These plants grow in wet meadows and bogs and at the edges of streams.

POISONOUS PARTS AND CONDITIONS OF POISONING: Water hemlock is one of the most poisonous plants in the Northwest. The whole plant is poisonous, but the root is most deadly.

Both the cut stalk and the root contain a yellow, oily liquid which has a strong smell. This liquid is especially poisonous.

Symptoms of poisoning include stomach pains, nausea, vomiting, weak and rapid pulse, and violent convulsions. There is some chance of survival by humans if free vomiting can be induced, followed by a laxative. Otherwise, the person will die in a few hours.

CAUTION: There are a number of other plants in the same family (Umbelliferae) which can be confused with this one: *Ligusticum scoticum, Angelica lucida, Angelica genuflexa, Conioselinum chinense.* If you are at all unsure which plant you have, don't even *touch* it!

PEA FAMILY

NOOTKA LUPINE
Lupinus nootkatensis
Kakoríki (Russian)

DESCRIPTION: The lupine is a perennial. It has a long taproot and stems up to 3½ feet tall, with many branches. The leaves are alternate, each having 6 to 8 leaflets which radiate out from a center. The leaflets are blunt-tipped, silky, and hairy. Flowers form at the top of the plant in dense clusters up to 10 inches long. They have up to 5 petals. Usually blue, flowers are often shaded pink or white and are occasionally all white. The fruit is black with a pea-pod shape, 1 to 1½ inches long.

HABITAT: Lupines like hillsides, open areas, and gravel bars.

POISONOUS PARTS AND CONDITIONS OF POISONING: The seeds in the pea-pod fruits are the most poisonous part of the plant, so must not be eaten.

Our references disagree on the poisonous properties of this plant. Some kinds of lupines are poisonous and some are not. In fact, it is said that the Aleuts used to gather the roots in the fall, scrape off the skin, and eat the inner portion. However, as lupines can contain poisons that cause fatal inflammation of the stomach and intestines, it is better not to try eating these plants at all.

Lupine *is* good for the soil, as it is one of the plants that add nitrogen to soil, making it richer for other plants.

WILD SWEET PEA
Hedysarum Mackenzii

DESCRIPTION: This plant is a perennial with a few erect stems up to 1½ feet tall, covered with tiny hairs. The leaves, round at both ends, have 9 to 17 small leaflets which are grayish and hairy underneath. The flowers are fragrant and very attractive, rose to violet-purple, arranged as in a sweet pea. The seed pods are flat, usually with 6 oval joints. Each joint contains 1 seed.

HABITAT: Grows in open, gravelly places and sandy river beds. This plant is not found as far south as Kodiak.

POISONOUS PARTS: This plant is reported to be poisonous, as there were cases where it apparently caused sickness in people. It is not to be confused with *Lathyrus maritimus*, or beach pea, which grows in our area and is edible.

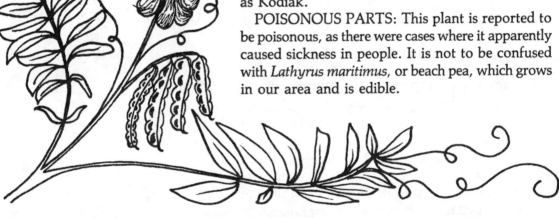

HINTS ON COOKING WILD EDIBLES

Very often it seems that the novice at foraging for wild edibles reaches a kind of plateau when it comes to the actual preparation of plants for the dinner table. The new plant enthusiast has learned to identify wild vegetables correctly, discovered where they are growing, harvested them, and brought them home. But what then?

Part of the purpose of this book is to make that next step easier. Specific recipes have been provided after each plant to help the cook get it on the table. The purpose of the lists that follow is to make that task even simpler. One can see at a glance how the plant in question can be used, and plan the meal accordingly.

In the introduction it was stated that wild edibles generally have a higher nutritional value than commercial vegetables. The best ways to get full benefits of that nutritional value are:

1) Eat plants as soon as possible after gathering.
2) Eat plants without cooking them.

Therefore, if used soon after picking in salads, the plants have maximum nutritional value — none of their vitamins have been destroyed by cooking.

However, not all wild plants can be put in salads. Then, too, the day's meal may call for a dish other than a salad. Many leaves of wild edibles can be cooked as potherbs, which means they are boiled or steamed and served like spinach. Some can be wilted with vinegar and bacon. Or they might replace vegetables like asparagus or green beans.

Some plants are quite tender and should not be overcooked. They can be steamed and eaten as they are, with butter, salt and pepper, or with a little vinegar. Others are coarser-textured and require more cooking time. Use a fork to check for doneness (fork should pierce plants easily).

Still another group of plants tends to be bitter or strong in flavor. Included here is the marsh marigold, which contains a poison that is destroyed by cooking. Place plants listed in this category in a small amount of water, cover, bring to a boil, then drain. Pour a fresh supply of boiling water into the pan and cook until plant is tender. Some plants, such as marsh marigold, may need the water changed twice to rid the plant of all undesirable properties.

Some wild plants are cooked in soup stocks and stews, or act as seasonings; others appear in purées and creamed soups (see recipes following).

A few Alaskan plants can be served as potato substitutes. The plants in this group are best gathered from autumn to spring because the starch stored in their roots is more abundant then. Also, during the cold months the roots are firm, but in summer, when the plant is using stored food, roots become mushy.

A few plants growing in the Kodiak area produce seeds that can be ground into flour. Before grinding, it is necessary to remove the husks and separate them from the seeds. Rub the seeds between two boards or two flat rocks, then pour them between two tin cans in a light breeze. The husks will blow away and the seeds will remain.

The seeds can be ground between two rocks to produce flour. However, as they need to be crushed as fine as possible, a knife-type kitchen blender or a hand flour mill might work better.

Seeds can also be boiled for cereal — add sweetener or bacon drippings to taste.

Two plants, dandelion and northern bedstraw, can be roasted and used as coffee substitutes.

Basic Purée Recipe

Simmer greens for 20 minutes, then press through a colander and mash, or use a meat grinder or blender. Add butter and seasonings.

Basic Cream Soup Recipe

Pour 1 quart rich cream in a pan; place over low heat. When cream starts to bubble, add 3 cups puréed greens; season with salt and pepper to taste.

Variations:
1) Reduce the amount of cream and add 1 can cream of mushroom soup.
2) Mince 1 onion, sauté in butter, and add.
3) Add ½ cup grated cheese.
4) Add 2 or 3 beaten egg yolks; blend well.
5) Add various seasonings, such as curry powder or paprika, to top soup.

Wild Salads
(Young plants best)

Alaska Spring Beauty
Beach Peas
Broad-leaved Plantain (very young)
Chickweed
Dandelion (leaves)
Fireweed (stems and leaves)
Goosetongue
Lambsquarter
Mountain Sorrel
Petrúshki
Rose (petals)
Roseroot
Salmonberry (blossoms)
Saxifrage
Seabeach Sandwort
Sourdock
Spearmint
Spreading Wood Fern (fiddleheads when very young)
Wild Chives
Wild Cucumber
Wild Violet
Willow (very young leaves)

COOKING HINTS

Wilted with Vinegar and Bacon

Chickweed
Dandelion
Goosetongue
Lambsquarter
Mountain Sorrel
Saxifrage
Sourdock

Blanched (boiled 1 minute), Drained, Sautéed

Fireweed (shoots)
Spreading Wood Fern (fiddleheads)

I. Cooked Greens
(Tender: Steam and don't overcook)

Alaska Spring Beauty (young plants)
Broad-leaved Plantain (young plants)
Chickweed
Clover (leaves of young plants)
Goosetongue
Lambsquarter
Mountain Sorrel
Sourdock

II. Cooked Greens
(Cook longer than Group I, but no more than necessary; use fork test)

Beach Peas
Clover (roots)
Dandelion (young roots)
Fern (fiddleheads)
Fireweed (shoots)
Nettle (boil or steam 5 to 15 minutes)
Wild Violet (leaves)

III. Cooked Greens
(May be bitter — bring to boil, drain, boil again)

Broad-leaved Plantain (older plants)
Dandelion (leaves of older plants)
Fireweed (young plants)
Marsh Marigold (boil in three changes of water)
Northern Bedstraw
Petrúshki
Sourdock (older plants)

Cooked in Soup Stocks and Stews; Used as Seasonings

Clover (roots)
Fireweed (peeled stems of older plants)
Horsetail (new shoots above rootstock)
Petrúshki
Póochki (leaves dried, burned, and powdered, or peeled stems)
Spearmint
Sweet Gale (leaves or berries, to season meat)
Wild Chives

Creamed Soups and Purées

Broad-leaved Plantain
Chickweed
Goosetongue
Lambsquarter
Marsh Marigold (boil in three changes of water)
Mountain Sorrel
Nettle
Póochki (peeled stems)
Saxifrage
Sourdock
Wild Cucumber

Edible Roots and Bark
(Source of starch or potato substitute)

Indian Rice
Roseroot (where abundant)
Silverweed
Willow (inner bark)
Yellow Pond Lily (boil twice)

Edible Seeds and Bark
(For flours and cereals)

Clover
Lambsquarter
Sourdock
Willow (inner bark)
Yellow Pond Lily

Coffee Substitutes

Dandelion (roots)
Northern Bedstraw (seeds and roots)

WILD GAME RECIPES

This chapter and the seafood chapter that follows contain some recipes that do not have wild plants as ingredients; these recipes are meant to be served with side dishes made with wild plants to create a whole meal that is a product of the natural environment. (For example, try any of the deer dishes with steamed lambsquarter or goosetongue.) These wild game and seafood recipes are typical of the Spruce Island/Kodiak area and its people.

Cooking Instructions for Seal

Seal meat contains a great deal of blood, so first boil it alone in lots of water. As it boils, keep taking off the scum that forms on top — just pour off the top part of the juice. Continue this process until the juice is clear. Cook at least 1 hour from the time it begins boiling. After the juice is clear, add vegetables if you wish.

Sea Lion Pot Roast
Rosemary Squartsoff

1 sea lion roast
Water
Salt
2 bay leaves
½ onion, chopped
2 cloves garlic
¼ pound bacon
Mustard

Cut meat into bite-size pieces and place in mixing bowl. Cover meat with water, salt, and bay leaves. Let soak about 16 hours, then drain and change water until it stays clear. Place meat in pot; cover with cold water. Bring to boil; remove blood as it forms. When blood stops coming to top, drain and reserve water and set meat aside. Sauté onion, garlic, and bacon over medium heat, stirring often, until tender. Add ½ to 1 cup reserved water to pan; add meat. Simmer, stirring often, from ½ to 1 hour, until done. Serve hot with mustard.

Boiled Sea Lion
Rosemary Squartsoff

1 quart chunks of sea lion meat
2 to 5 potatoes
Mustard

Place sea lion meat in cold water to cover. Bring water to boil. As blood boils to top, remove. When water clears, drain it off. Cover meat with hot water. Add salt to taste; boil until tender but still chewy, from ½ to 1 hour depending on age of animal. Add potatoes and boil 25 minutes more. Serve with hot mustard.

Aleut Pilaf
Rosemary Squartsoff

⅓ cup rice, washed
3 tablespoons butter
½ cup stewed tomatoes
½ cup cooked sea lion, diced
Duck liver, diced
Chicken stock
Salt
Pepper
Butter (optional)

Cook rice in well-salted boiling water. Drain. Pour hot water over rice and rinse well. Melt butter in omelette pan, add rice, and cook 3 minutes. Add tomatoes, sea lion, and duck liver. Sauté, with enough chicken stock to moisten, for 5 minutes. Season with salt and pepper. Add more butter if desired. Serves 4.

Deer Swiss Steaks
Angeline Anderson

6 medium deer steaks
Salt
Pepper
Flour
Cooking oil
1 onion, diced
2 stalks celery, diced
2 cups water
3 tablespoons dried petrúshki (beach lovage)
1 small can mushrooms
Worcestershire sauce
Flour

Season steaks with salt and pepper and roll in flour. Brown on both sides in oil. Remove to roasting pan. Brown onion and celery in oil until tender. Add water, mushrooms, petrúshki, browned onion and celery, and Worcestershire. Thicken gravy with flour. Pour over steaks and simmer 45 minutes.

Deer Burgers
Sasha Smith

1 deer heart
1 small onion
Salt
Pepper

Grind deer heart and onion; mix well. Add salt and pepper to taste. Form into patties and fry until brown on one side. Turn and brown. Serve with gravy.

Deer Roast
Georgia Smith

3 to 4-pound deer roast
Cooking oil
1 package dry onion soup mix
1 can cream of mushroom soup
2 tablespoons Worcestershire sauce

Brown roast on each side in oil in electric skillet. Combine soups and Worcestershire sauce. Pour over roast; cover and reduce heat. Simmer 2½ to 3 hours. Thin with water if sauce thickens.

Deer Bean Boil
Rosemary Squartsoff

2 pounds pinto beans
2 pounds red beans
15-ounce can stewed tomatoes
15-ounce can tomato sauce
1 pound smoked hocks
2 pounds deer meat, cut into 2-inch squares
2 chorizo links
1 tablespoon salt
Chili pepper
1 large onion, diced
4 cloves garlic, diced (optional)

Wash beans and place in pot. Bring to boil. Crush stewed tomatoes and add to pot along with tomato sauce and hocks. Bring to second boil, reduce heat, and simmer until beans are almost done (about 1 hour). Add deer meat and rest of ingredients (except onion and garlic) and cook until meat is tender (about 1½ hours). Onion and garlic should be added during the last half-hour of cooking time.

Baked Deer Spareribs
Georgia Smith

2 sides spareribs
1 teaspoon onion salt
½ teaspoon celery salt
½ teaspoon garlic salt
1 teaspoon sage
½ teaspoon marjoram
Salt
Pepper
About 1 cup water
4 medium potatoes
10 small carrots
2 medium onions

Cut spareribs into serving size pieces. Mix seasonings; rub into meat. Place spareribs in roasting pan. Add 1 cup water. Cover; bake at 350 degrees for 1 hour. Remove lid for last 15 to 20 minutes if you want it to brown more. Cut potatoes, carrots, and onions in half; arrange around meat. Cover and bake 1 hour longer. Add more water if needed.

Elk Mulligan
Rosemary Squartsoff

1 pound elk meat
2 large potatoes
2 carrots
1 stalk celery
1 rutabaga
1 medium onion
¼ pound butter (optional)
½ cup rice
½ cup wild carrot leaves (optional)
½ cup petrúshki
Salt
Pepper

Cut elk meat into 2-inch cubes. Place in pot, cover with water, and boil until almost done (about 1½ hours). Cut vegetables into fairly small pieces and add with rice and butter during the last half-hour. Chop carrot leaves and petrúshki and add during last 20 minutes or so of cooking time. Season with salt and pepper.

Duck Soup
Sasha Smith

2 medium ducks, cut up
2 potatoes, diced
3 carrots, diced
½ cup rice
Salt
Pepper

Cover the duck pieces with water and simmer for 2 hours on slow heat. Add potatoes, carrots and rice; cook for 40 minutes longer. Season with salt and pepper.

SEAFOOD RECIPES

Most of the recipes in this chapter and in the wild game chapter are meant to complete a meal prepared with wild plants. Serve one of the halibut dishes with Spruce Island Weed Salad (seabeach sandwort under Pink Family), for example. Other recipes include wild plants as ingredients in the specific dish. All of the recipes, whether made with wild plants or intended to complement them, derive from the natural environment. They are favorite regional recipes.

Shrimp de Jonghe
Fran Kelso

1 cup butter, melted
2 cloves garlic, minced
⅓ cup chopped parsley or petrúshki
½ teaspoon paprika
Dash cayenne
⅔ cup cooking sherry
2 cups soft bread crumbs
5 to 6 cups (4 pounds in shell) cleaned cooked shrimp

To melted butter, add garlic, parsley, paprika, cayenne, and sherry; mix. Add bread crumbs; toss. Place shrimp in 11 by 7 by 1½-inch baking dish. Spoon the butter mixture over shrimp. Bake in slow oven (325 degrees) for 20 to 25 minutes or until crumbs brown. Sprinkle with additional chopped parsley before serving.

Shrimp Salad
Fran Kelso

1 cup tomato soup
2 (8-ounce) packages cream cheese
1 package gelatin dissolved in ¼ cup water
½ cup mayonnaise
1½ cup (each) chopped celery, onion, and green pepper
2 small cans or 1 large can small shrimp

Heat tomato soup with cream cheese and whip smooth. Add dissolved gelatin and let cool. Add rest of ingredients. Chill.
Variation: Chicken can be substituted for shrimp.

Crab Meat Rolls
Clay Ferris

½ cup tomato juice
1 egg, beaten
1 cup dry bread crumbs
⅛ teaspoon salt
Dash pepper
½ teaspoon chopped petrúshki
½ teaspoon chopped celery leaves
6-ounce can crab meat, flaked
12 slices bacon, cut in half

Mix tomato juice and egg. Add bread crumbs, salt and pepper, petrúshki, celery leaves, and crab meat. Mix thoroughly; roll into finger lengths; wrap each with ½ slice bacon. Fasten with toothpicks. Broil, turning frequently to brown evenly. Makes 2 dozen.

Clay's Crab Balls
Clay Ferris

1 pound crab meat, cooked or canned
1 egg, well-beaten
¼ cup cracker crumbs

Keep large chunks of crab whole; reserve. Flake the small pieces of crab meat and mix with beaten egg and cracker crumbs. Shape into 1-inch balls. Chill.

Batter:

1 egg, slightly beaten
½ cup water
½ cup flour
2 cups peanut oil

Mix egg, water, and flour. Dip the crab meat chunks and the balls into the batter and fry them in hot oil until brown. Drain on a paper towel. Serve with sauce (below). Serves 4.

Sauce:

½ cup hot beef bouillon
2 tablespoons soy sauce
1 teaspoon sugar
1 teaspoon prepared horseradish
½ teaspoon blended salad herbs

Mix all ingredients together and serve.

Seaside Sandwich
Given to Fran Kelso by Sally Ramaglia

6 ounces grated cheese
6½-ounce can king crab meat, flaked
1 tablespoon chopped onion
½ teaspoon chili powder
Dash salt
2 teaspoons lemon juice
4 English muffins, cut in halves

Mix cheese, crab meat, onion, chili powder, salt, and lemon juice. Spread crab mixture over English muffins. Heat under broiler until sandwich tops are golden brown. Serve hot. Makes 8 open-face sandwiches.

Spaghetti a la King Crab
Mary Craig

2 (7½-ounce) cans crab meat or 1 pound fresh or frozen king crab
2 tablespoons olive oil
½ cup butter
4 cloves garlic, minced
1 bunch green onions, sliced
2 medium tomatoes, diced
½ cup parsley or petrúshki
2 tablespoons lemon juice
½ teaspoon Italian seasoning
1 teaspoon salt
1 pound spaghetti
Parmesan croutons
Grated parmesan cheese

Drain and slice canned crab or defrost, drain and slice frozen crab. Heat oil and butter. Add garlic and sauté gently. Add crab, green onions, tomatoes, parsley, lemon juice, Italian seasoning, and salt. Heat gently 8 to 10 minutes. Meanwhile, cook spaghetti in boiling, salted water just until tender. Drain spaghetti. Toss with king crab mixture and Parmesan croutons. Pass additional grated Parmesan cheese. Makes 6 servings.

King Crab Soufflé
Given to Fran Kelso by Sally Ramaglia

8 slices bread
2 cups crab meat
½ cup mayonnaise
1 onion, chopped
1 green pepper, chopped
1 cup chopped celery
4 eggs
3 cups milk
1 can mushroom soup
Grated cheese

Dice half the bread and put pieces into large baking dish. Mix crab meat, mayonnaise, onion, green pepper, and celery; spread over bread pieces. Trim crusts from remaining 4 slices bread and place over crab mixture. Mix eggs and milk together and pour over bread. Place in refrigerator overnight. Bake in 325 degree oven for 15 minutes. Take from oven and spoon soup over the top; sprinkle with grated cheese and paprika. Bake 1½ hours in 325 degree oven. Yields 8 servings.

Cat's Crab Dip
Cathy Ray Klinkert

1 (8-ounce) package cream cheese
¾ cup crab meat
3 to 4 tablespoons minced onion
1 to 2 cloves minced garlic
Corn chips or tortillas

In a heavy saucepan, mix all ingredients except corn chips and heat slowly, stirring constantly to keep cheese from sticking or burning. Serve with corn chips or pieces of corn tortilla that have been fried in oil until crisp, then lightly salted.

Variations: Salmon, shrimp, or halibut may be substituted for the crab. This dish is an excellent way to use canned salmon.

Ooeduck (Baidarka or Gumboot) Soup
Angeline Anderson

Boil cut-up cabbage, potatoes, carrots, celery and rice in water.

In another pot, boil water; put in gumboots for 1 minute. Remove gumboots from water; clean and take out milk (yellow and white part). Chop into small pieces. Fry in butter with onions. Add some petrúshki, if available, for seasoning.

Stir flour into vegetables to make a gravy, or make a white sauce and pour it into vegetables. Add gumboots and onions to thickened vegetables. Heat through and serve.

Gumboot Rice in Gravy
Angeline Anderson

Clean gumboots and cut into small pieces. Fry onions, celery, green pepper, and 1 minced clove garlic in oil. Add gumboots last. Make a gravy with flour. Add seasonings of your choice, Worcestershire sauce, and salt to taste. Add gravy coloring. Serve over cooked rice. Good!

Cockle Clam Salad
Sasha Smith

3 cups cockle clams
1 egg, boiled and diced
2 tablespoons diced onion
Lettuce
Mayonnaise

Cook clams for a few minutes, then put through a grinder. Add egg, onion, lettuce and mayonnaise. Stir.

Clam Chowder
Sasha Smithy

4 cups fresh clams, minced or whole
2 potatoes, diced
1 onion, diced
3 slices bacon, diced and fried
Salt
1 cup milk

Combine all ingredients except milk in a saucepan. Cover with water and bring to a boil. Boil 20 minutes and add milk. Do not boil after adding milk; heat well and serve.

Clam Fritters
Sasha Smith

1 small onion
4 cups fresh clams
1 egg
4 tablespoons flour
Salt
Pepper
Cracker or bread crumbs (optional)

Put onion and clams through grinder. Add egg, flour, salt, pepper, and cracker crumbs, if used. Drop by tablespoons in hot fat. When brown, turn (usually takes 3 or 4 minutes on each side). Serve with gravy.

Nell's Favorite Scallop Recipe
Nell Tsacrios

Open shell. Remove meat. Dip in lemon juice and soy sauce. Eat.

Scallop Broth
Ozzie Walters

A tasty, invigorating drink for a fisherman on a cold night.

Take a handful of fresh-shucked scallops and poach them in 1 can of beer for 6 to 8 minutes. Throw the scallops away and drink the broth. (If you want to keep the scallops, chill them and serve them cold in a salad.)

Danny's Spicy Steamed Mussels
Danny Konigsberg

4 dozen large mussels in shells
1 crushed bay leaf
1 small onion, chopped
3 tablespoons chopped petrúshki
1 stalk póochki, peeled and minced
1 tablespoon wine vinegar
4 ounces butter
⅛ teaspoon cayenne
¼ teaspoon Worcestershire sauce
Salt

Scrub mussels; pull off any sea animals. Soak in sea water for 3 days, changing water daily, to clean.

Place mussels in large saucepan over medium heat; add bay leaf, onion, petrúshki and póochki. Cover tightly and steam for 3 minutes, or until shells begin to open. Drain broth from saucepan and save. Cover mussels to keep hot.

Strain broth into a smaller pan and add wine vinegar, butter, cayenne, Worcestershire sauce, and a little salt. Heat over medium flame to boiling point, but do not boil. Meanwhile, open mussels and remove top shell; take out dark, hairy beard. Serve mussels on half shells with seasoned butter sauce on the side. Dip each mussel in sauce and eat.

Variation: Use 1 teaspoon lemon juice in place of vinegar in butter sauce.

Fish Casserole
Mary Abrams

2 pounds halibut, cod, bass, shrimp, scallops, or combination of other fish
1 onion, chopped
3 stalks celery, chopped
½ green pepper, chopped
2 cups thick white sauce
⅓ bar easily melted cheese
Buttered crumbs
Paprika

Simmer fish until it can be easily broken apart. Meanwhile, sauté onion, celery, and green pepper. Make white sauce; add most of cheese and cook until melted. Combine fish, sautéed vegetables, and sauce and pour into buttered casserole. Top with buttered crumbs, grated remaining cheese, and paprika. Bake at 375 degrees for 30 to 40 minutes.

Codfish Bacon Bake
Rosemary Squartsoff

Clean and fillet fish; lay out meat side up in baking pan. Sprinkle all over with soy sauce, then dill pickle liquid. Marinate 4 hours. Cover fish with bacon and pan with foil. Heat oven to 450 degrees; bake 20 to 25 minutes depending on thickness of fish. Take off foil and cook 5 to 10 minutes longer, until bacon browns.

French Fried Sea Bass
Georgia Smith

2 pounds sea bass fillets, cut in 1-inch pieces
Very thin pancake batter

Dip pieces of fish in batter. Fry in hot fat and drain on paper towels.

Zesty Lemon Roll-Ups
Georgia Smith

⅓ cup butter
⅓ cup lemon juice
2 teaspoons chicken bouillon granules or 2 bouillon cubes
2 teaspoons Worcestershire sauce
1 cup cooked rice
5 ounces frozen chopped broccoli, thawed
¼ cup chopped petrúshki
1 cup shredded cheddar cheese
8 white fish fillets (halibut, cod, etc.)
Paprika

Preheat oven to 375 degrees. In small saucepan, melt butter; add lemon juice, bouillon, and Worcestershire sauce. Heat slowly until bouillon dissolves; set aside. In medium bowl, combine rice, broccoli, petrúshki, cheese, and ¼ cup lemon butter sauce; mix well. Divide broccoli mixture equally among fillets. Roll up and place seam-side down in shallow baking dish. Pour remaining sauce over roll-ups. Bake 25 minutes, or until fish flakes with fork. Spoon sauce over individual servings; garnish with paprika.

Poached Halibut
Ozzie Walters

Water
Soy sauce
White wine

In a wide pan, make a poaching liquid of any combination of the above ingredients.

Onions
Halibut fillets
Paprika or cayenne

Chop or ring onions and add to poaching liquid; cook 3 minutes. Place skinned fillets of halibut in the poaching liquid so they are not too crowded. Poach 10 minutes per inch of thickness of fillets. Garnish with paprika or cayenne.

Sauce: Tart berry jam or sauce goes well with the halibut. Try a sweet and sour sauce composed of salmonberry jam, a little sugar, and a little vinegar.

Baked Halibut with Mushrooms
Georgia Smith

1 cup mushrooms, cooked and drained
½ cup chopped onion
3 tablespoons diced green pepper
1 teaspoon lemon juice
1 (8-ounce) can tomato sauce
1 teaspoon salt
½ teaspoon sugar
⅛ teaspoon black pepper
1½ pounds halibut steaks

Preheat oven to 350 degrees. Place mushrooms, onion, green pepper, lemon juice, tomato sauce, salt, sugar, and black pepper in medium saucepan. Bring to boil, stirring constantly. Reduce heat and simmer, uncovered, for 2 to 3 minutes. Spoon half of mixture into lightly greased 8 by 12-inch pan. Top with fish. Cover with remaining mushroom mixture. Bake, uncovered, until fish flakes — about 20 minutes. Arrange fish on heated platter. Spoon mushroom sauce over fish.

Fish Chowder with Petrúshki
Sasha Smith

2 to 3 pounds halibut or salmon, cubed
2 potatoes, diced
1 onion, diced
½ cup rice
½ cup chopped petrúshki, fresh or dried
Salt
Pepper

Put halibut or salmon, potatoes, onion and rice in saucepan. Cover with water and bring to a boil. Cook 10 minutes and add petrúshki; season with salt and pepper. Boil 10 minutes longer.

Variations: Halibut or salmon heads can be used in this recipe.

Nellie's Halibut Supreme
Nell Tsacrios

About ½ stick butter
1 large clove garlic, pressed
Halibut, cut in 2 by 3-inch pieces
White flour
½ cup half-and-half
½ cup Chablis
1 tablespoon chopped parsley (preferably fresh) or petrúshki
Salt
Sliced almonds

In a 12-inch skillet, sauté garlic in butter. Add enough fish to nearly fill skillet. Remove fish when browned. Add more butter, if necessary, to make at least 2 tablespoons in pan. Add flour to make sauce. Thin with half-and-half and Chablis. Simmer until thick; add chopped parsley and salt to taste. Put fish back in pan; cover with sauce. Simmer on as low heat as possible until fish is cooked through (approximately 5 to 7 minutes). Sprinkle with sliced almonds before serving.

Fish in French Batter
Fran Kelso

1 egg
1 cup flour
½ teaspoon salt
1 teaspoon salad oil
½ cup water
1 teaspoon brandy (optional)
Fish fillets (halibut is excellent)
Tartar sauce

Beat egg well. Beat in other ingredients, adding water slowly until desired consistency is reached. Batter should be smooth and quite thick. If a lighter batter is preferred, beat yolk and white separately. Fold in the stiffly beaten egg white after the other ingredients have been well blended.

Cover fish fillets thoroughly with batter and fry in hot oil. Serve with tartar sauce.

Sea Soup
Betty Blackshear

1 cup chopped onion
1 tablespoon margarine
1 can tomatoes, chopped
1 teaspoon minced garlic
1½ cups (2 6-ounce cans) chopped, cooked clams
2-inch piece orange peel
¼ teaspoon dried thyme leaves
⅛ teaspoon pepper
½ cup chopped fresh parsley or 1 tablespoon dried parsley
1 pound fish fillets, cut in 2-inch chunks

In a medium saucepan, cook onion in margarine until tender. Add drained tomatoes (reserve liquid) and garlic; cook about 5 minutes. Add liquid from clams, tomato juice, orange peel, thyme, pepper, and half of the parsley. Bring to boil and simmer 15 minutes. Increase heat and add clams and fish while mixture is boiling. Cook 5 minutes or until fish is cooked. Top with remaining parsley.

Fish Head Soup
Rosemary Squartsoff

Fish heads
1 onion, chopped
2 potatoes, chopped
⅓ cup rice
Petrúshki
Salt
Pepper

Boil fish heads, onion, potatoes, and rice for 20 minutes. When done, add some petrúshki, season with salt and pepper, and boil 10 minutes longer. Serve.

Danny's Salmon Loaf
Daniel Konigsberg

2 cups cooked salmon
⅔ cup evaporated milk
2 cups soft bread crumbs
1 egg, well beaten
1 teaspoon petrúshki
2 teaspoons minced onion
6 teaspoons minced green pepper
2 teaspoons minced celery
½ teaspoon salt
¼ teaspoon poultry seasoning
6 ounces chopped green chiles
Cheddar cheese, grated

Heat oven to 375 degrees. Mix all ingredients except cheese. Spread in greased baking dish; sprinkle with grated cheddar cheese. Bake until center is firm (about 45 minutes).

Salmon and Nettles with Sunflower Seeds

2 cups nettles
1 tablespoon lemon juice
3 tablespoons butter
3 tablespoons whole wheat flour
3 cups milk, scalded
½ cup grated Parmesan cheese
½ teaspoon cayenne pepper
⅛ teaspoon nutmeg
1 teaspoon ground mustard seeds
2 cups cooked salmon
4 tablespoons slivered almonds, toasted

Boil nettles 5 to 10 minutes. Drain and put in bottom of buttered casserole. Sprinkle with lemon juice and set aside.

In a saucepan, melt the butter and add the whole wheat flour. Stirring with a wire whisk, gradually add the milk. Cook over low heat, stirring constantly, until mixture has thickened. Add Parmesan cheese, cayenne pepper, nutmeg, and mustard seeds. Let cook for 1 minute and add the salmon. Pour this mixture over the nettles and bake for 15 minutes at 450 degrees. Sprinkle with almonds and bake 5 minutes more.

Salmon Perok
Sasha Smith

Pastry for 2 double-crust pies
2 cups rice, cooked with 2 tablespoons butter
Salt
Pepper
1 salmon, filleted
1 large onion, chopped

Use half of pastry to line a 9 by 13-inch pan. Place half the cooked rice over the pastry. Sprinkle lightly with salt and pepper. Add a layer of salmon, skin side down. Add more salt and pepper; place onion on salmon. Top with another layer of rice. Season again with salt and pepper. Fold top crust under edge of bottom crust to seal. Make slits in top crust for steam to escape. Bake at 400 degrees for 1 hour or until brown on top.

Salmon Ring with Cheese Sauce
Chris Abell

1 egg
1 cup cooked salmon
½ cup chopped onion
½ cup sharp cheddar cheese
1 teaspoon celery salt
¼ teaspoon pepper
Parsley
2 cups buttermilk baking mix
½ cup cold water

Heat oven to 375 degrees. Beat egg slightly; set aside 2 tablespoons. Stir salmon, onion, cheese, celery salt, pepper, and parsley into remaining egg. Stir baking mix and water into a soft dough; knead 5 times on floured, cloth-covered board. Roll into rectangle, 15 by 10 inches. Spread with salmon mixture.

Roll up, beginning at the long side. With sealed edge down, shape into ring on greased baking sheet; pinch ends together. With scissors or knife, make small cuts on top of ring about every 1½ inches. Brush with remaining egg. Bake 25 to 30 minutes. Serve with hot cheese sauce.

Cheese Sauce:
¼ cup butter or margarine
¼ cup buttermilk baking mix
¼ teaspoon *each* salt and pepper
2 cups milk
1 cup shredded cheddar cheese

Melt butter over low heat. Blend in baking mix, salt and pepper. Cook over low heat, stirring, until smooth and bubbly. Remove from heat; stir in milk. Heat to boiling, stirring constantly. Boil and stir 1 minute. Stir in cheese until melted.

Canned Salmon with Garlic
Sasha Smith

Heat contents of a 1-pound can of salmon in a baking dish in oven. Sprinkle with 1 clove of garlic, minced. Serve.

Fisharoni Surprise
Georgia Smith

2 cups cooked macaroni
1 medium onion, chopped and sautéed in butter
1-pound can of salmon, flaked
1 medium-size can baked beans
Salt
Pepper
1 small can (10½-ounce) tomato soup
2 tablespoons butter or margarine

Place half of cooked macaroni in bottom of greased casserole. Combine onion and salmon; place half of this mixture over macaroni in baking dish. Add baked beans, spreading evenly. Add remaining fish and rest of macaroni. Sprinkle with salt and pepper. Top with can of tomato soup. Dot with butter. Bake at 350 degrees for 20 to 30 minutes. Serves 4 to 6.

Barbecued Fish Fillets
Georgia Smith

6 tablespoons butter
½ cup diced onion
1½ pounds salmon or trout fillets
Salt
Pepper
½ cup catsup
2 teaspoons sugar
⅓ cup lemon juice
2 teaspoons Worcestershire sauce
¼ cup water

Preheat skillet; add 2 tablespoons butter. When melted, add onions and sauté. Remove onions. Add remaining butter. Cut fish fillets into serving-size portions. Brown lightly in the butter, turning carefully with spatula. Spread sautéed onions over fish. Season with salt and pepper. Combine remaining ingredients; pour over fish. Simmer 20 minutes or until fish flakes easily.

Easy Salmon Bake
Fran Kelso

1 red salmon, filleted
Lemon juice
Mayonnaise
Dill weed, crushed

Place filleted salmon, skin side down, in a baking dish. Sprinkle liberally with lemon juice and spread with a thin layer of mayonnaise. Sprinkle crushed dill weed on top. Bake at 350 degrees for 20 to 30 minutes, until cooked through.

Deep Dish Salmon-Wild Rice-Nettle Pie
Stacy Studebaker

Silver salmon
Wild rice
Nettles
Mushrooms
Onions
Seasonings
Butter
Pastry

Stacy combines these ingredients according to her mood and what she has on hand.

Octopus Salad
Jim Tsacrios

Kill octopus. Turn inside out and remove innards. Beat carcass about 75 to 80 times, *hard*, on rocks at water's edge. After every 10 smacks, rub carcass on a rough rock until it foams; then rinse in sea water. Repeat process until carcass stops foaming.

Cut into pieces small enough to fit in a pot, and cover with water. Boil for 20 to 30 minutes. Slice into thin pieces; very thin if octopus is fairly large. Put in marinade of:

1 part vinegar
2 parts oil
2 or 3 onions, sliced in small moons

Marinate and chill for at least 24 hours, then serve.

SPRUCE ISLAND TEA RECIPES

A very positive thing happened to Plants Class (the Ouzinkie Botanical Society) while I was away traveling for two months in the winter of 1982. The class members continued working on projects of their own, simply because they were interested in what they were learning.

One of the things they did was make their own teas. They pooled resources and gathered together ingredients from collections of dried herbs each of them had on hand. They obtained empty tea bags for make-your-own blends from Nichols Garden Nursery, 1790 North Pacific Highway, Albany, Oregon 97321. It became Rosemary's task to determine what quantities of each herb they would need. The herbs were crushed and mixed in a blender before the tea bags were filled.

A set of these tasty teas and recipes was waiting for me when I returned home, and now I'll pass them on to you. Perhaps you'll think of your own blends to add to these. Meanwhile, for your sipping pleasure, here are Rosemary's herbal blends:

Tea Recipes
by Rosemary Squartsoff
and Plants Class

1. Chamomile or pineapple weed, ginger, orange peel (equal parts of each)
2. Yarrow, rose petals, mint or spearmint (equal parts of each)
3. Rose petals, lavender, chamomile or pineapple weed (equal parts of each)
4. Rose petals, marsh fivefinger flowers, mint or spearmint, cinnamon (equal parts of first 3; pinch of cinnamon)
5. Rose petals, mint or spearmint, goldenrod leaves (equal parts of each)
6. Mint or spearmint, sourdock root, petrúshki (equal parts of each)
7. Rose petals (½ ounce) and black tea (2 ounces)
8. Black tea (4 ounces), bay leaves (½ ounce), cinnamon (½ ounce), orange peel (½ ounce)
9. Rose petals (½ ounce), chamomile or pineapple weed (2 ounces), mint or spearmint (½ ounce)
10. Sage (½ ounce), cinnamon (¼ ounce), yarrow (2 ounces), mint or spearmint (1 ounce)
11. Comfrey (½ ounce), mint or spearmint (1 ounce), cinnamon (½ ounce)

12. Sage (½ ounce), orange peel (1 ounce), cinnamon (1 ounce)
13. Orange peel (1 ounce), Labrador tea (½ ounce), cinnamon (½ ounce)
14. Comfrey (½ ounce), mint or spearmint (½ ounce), orange peel (1 ounce)
15. Black tea (2 ounces), orange peel (1 ounce), mint or spearmint (½ ounce)

We hope you'll try — and enjoy — these blends. One final note: When you brew your tea, use a ceramic, enamel, or glass container — never a metal one. Cover the container while the tea steeps. Five minutes is about right for steeping most herb teas. For sweetener, use a small amount of honey — too much will cover the delicate flavors.

Now it's your turn. Use the space below to write down your own herbal tea recipes:

DYES FROM WILD PLANTS

Since ancient times, people all over the world have used plants to produce dyes for fibers, woven textiles, and leathers. Early Alaskan people were no exception. Yarns for clothing or ceremonial garments were dyed with plants gathered for that purpose. Dyed bark and grass fiber brightened baskets. Early fishermen in Alaska colored their nets with homemade dyes.

Today, we can imitate the ancients and use natural dyes made from the plants growing around us to produce many pleasing yarn colors.

A great variety of plants in the Kodiak area can be utilized to produce dyes. In this chapter we include a list of those we have researched, with recipes for making dyes from them. We have tried some of these dye recipes; others remain to be tested in future Plants Class (Ouzinkie Botanical Society) sessions. There are other Alaskan dye-producing plants and recipe variations not included here, but the ones that follow will give the beginning student a good introductory sample. Use these recipes as guidelines for experimentation.

The Plants

Some plants can be dried and used later for dyeing; others are best when used fresh. Blossoms picked for dyeing, for example, sometimes lose their color when they are dried. Most plants suitable for dyeing produce best results when picked in the late summer or autumn, or at the peak of their growing season.

The usual way to prepare plants for the dye bath is to crush or chop them, cover with water, and let stand overnight. (As a general rule, use a 2-gallon container of plant material to produce 4 gallons of dye bath.) The next day, boil plants 30 minutes to 2 hours, depending on the shade desired. Strain out the plants and add enough water to make 4 gallons; mix well and heat.

Two Processes

Usually, dyeing with plants calls for two processes. The first process is "mordanting"; the second is the dyeing itself. A mordant is a substance used to fix the color — make it permanent. Common mordants are alum, chrome, iron and tin. Other mordants include vinegar or acetic acid, ammonia or urine, blue vitriol, caustic soda or sodium hydroxide, lime, tannic acid, and cream of tartar.

Mordanting and dyeing can be done at the same time, but a clearer color will result if they are done separately. Be sure to allow enough time for each process.

The Yarn

Any natural fiber — *not* synthetic — can be dyed. Sometimes dyeing different fibers requires differing procedures. The directions given here are for wool yarn.

Use clean yarn. If the yarn is dirty or greasy, wash it in mild soap. If yarn is dry, wet it thoroughly before mordanting or dyeing, and squeeze — don't wring — to remove excess water.

For either mordanting or dyeing, *do not let wool yarn boil* — keep it simmering. Be careful to increase or decrease water temperature gradually. Drastic temperature changes cause wool to be "shocked" and result in shrunken, felted or matted yarn.

Let water cool slowly before removing yarn from either mordant or dye bath. When yarn is cool enough to handle, lift it out and rinse it in water that's the same temperature as the mordant or dye bath you took it from. Let the yarn dry slowly.

Equipment

The pot used for mordanting and dyeing should be copper, stainless steel, or enamel. Have available a cooking thermometer, large heat-resistant glass measuring cups, plastic spoons for measuring mordants, glass rods or wooden sticks for stirring (if sticks are wooden, use a separate one for each color), and plastic or enamel containers for rinsing yarn.

Water

For best results, use soft water for dyeing; rain water is best. Hard water contains dissolved mineral salts. Because of the minerals, dye colors will be less bright and clear. However, if only hard water is available, the addition of a little white vinegar will help soften it.

Crystals and Salts

If a recipe calls for copperas crystals (iron or ferrous sulfate), purchase the crystals in a drug store or from a chemical company.

Some recipes call for Glauber's salts. Although dyeing can be done without these, they are suggested because they make the dye color uniform. If Glauber's salts are used, it is not necessary to stir the dye bath. The salts can be purchased at a chemical company or a drug store.

Alum Mordant

When using an alum mordant, dissolve 3 ounces alum and 1 ounce cream of tartar in a small amount of water, then add it to 4 gallons lukewarm water. (This quantity is sufficient for 1 pound of yarn.) Then add wet yarn, slowly heat to just below boiling, and simmer 1 hour. Turn the yarn occasionally. Cool; remove from mordant bath.

If you don't plan on dyeing right away, let yarn dry slowly. It might take a few days for the yarn to dry completely. Before putting it in the dye bath, moisten yarn and squeeze gently to remove excess water.

Chrome Mordant

To use a chrome mordant, dissolve ½ ounce chrome in a small amount of boiling water. Pour into 4 gallons of hot water, add 1 pound yarn, and simmer, covered, 1 hour. Cool to lukewarm (see "The Yarn," preceeding) and rinse. NOTE: Whenever using chrome mordant, dye immediately after mordanting. Squeeze yarn gently to remove excess water before putting in dye bath.

Copperas Crystals
(Iron or Ferrous Sulfate) Mordants

These mordants color fibers green; they improve green or blue tones in the dyed yarn.

Add 2 ounces iron vitriol to 3 gallons warm water. Immerse 1 pound yarn and simmer 1 hour for dark green, a shorter time for lighter shades.

Dyeing Procedure

For each pound (dry weight) of wool yarn, use 4 gallons dye bath. Immerse moist yarn when the bath is lukewarm. Then heat slowly to a simmer and let simmer for ½ hour, occasionally stirring gently. If the water level gets low: remove yarn, add hot water, mix well, and return yarn to dye bath.

If dyeing and mordanting are done together, dissolve the mordant in the warm dye bath before adding the yarn (add cream of tartar, if used, after yarn has been in the bath 30 minutes). Leave yarn in the bath for 1 hour.

When dyeing time is completed, rinse the yarn several times in water of gradually decreasing temperatures until rinse water is clear. Squeeze out excess moisture and dry yarn in a cool place.

DYES FROM LICHENS

Lichens are one of nature's most mysterious plant forms. These tiny organisms are unique because they are both algae and fungi. (Algae are plants with chlorophyll but no vascular system, such as stem or leaf veins, for carrying essential fluids. Fungi are plants that have no chlorophyll; mushrooms are an example.) Both parts of the lichen cooperate so that the total plant flourishes in what is called a symbiotic union. Because of this alga-fungus cooperation, the lichen is one of the most adaptable and widely distributed plants on earth; there are thousands of species of lichens.

For centuries, lichens have had a variety of uses. Medicinally, they have been sought for treatment of skin ailments, for making antibiotics, and for curing tuberculosis. They are also used in making perfume. However, one of their most interesting uses is as a dye for yarn.

One of the nicest things about dyeing with lichens is that they require no mordanting, although mordants can be added if desired. Lichens also give wool yarn a pleasant smell, and it is said that they make wool mothproof.

The Plants Class has experimented with five different types of lichens, using no mordants. From the pale-green hairlike lichen known as old man's beard, in the genus *Usnea*, we obtained a tan color. From a member of the genus *Cladonia* we came up with a rich medium-brown shade. From a genus *Parmelia* lichen, nicknamed by Stacy Studebaker the "bird-dropping lichen" because of its appearance, we got a deep brown color. From two other *Parmelia* lichens we obtained a dark tan and a gold.

We urge you to experiment. Just be sure, as you collect, to gather lichens that are all the same type. We also suggest that, for lichens and all other plants used in dyeing, you save a small portion of the plant along with a little piece of the dyed yarn as a record of the color you obtained.

Lichen Dye

Gather about 2 gallons of the type of lichen you wish to try. (You might find it easier to gather the lichens after a rain.) Soak lichens overnight in water to cover. Boil in the same water for 1 hour, strain, add enough water to make 4 gallons, add 1 pound yarn, and simmer for ½ hour or longer. (Adjust amounts of water and yarn to strength of dye material.)

No mordant: Color will vary, depending on type of lichen used.
With alum mordant: Yellow tan.
With $\frac{1}{8}$ ounce chrome and $\frac{1}{16}$ ounce vinegar or acetic acid (see chrome mordant directions): Rose tan.

NOTE: Mordanted yarn shades might vary with different kinds of lichens.

DYES FROM SMALL PLANTS

Broad-leaved Plantain *(Plantago major)*:

According to references, plantain produces colors from green to yellow brown when mordanted. We do not have the recipe — feel free to experiment.

Goldenrod
(Solidago lepida, S. multiradiata):

Blossoms

Pick when first in bloom. Gather 1 to 1½ pounds flowers and cover with water. Let stand in water overnight. Boil 1 hour or longer. Strain, add enough water to make 4 gallons, add 1 pound yarn, and simmer 1 hour.

With alum mordant: Yellowish tan.
With $\frac{1}{8}$ ounce chrome and $\frac{1}{8}$ ounce vinegar or acetic acid (see chrome mordant directions): Old gold.

Whole plant

Gather 4 gallons goldenrod plants in late summer or autumn. Cut into 1 to 3-inch pieces, cover with water, and boil 2 hours. Add more water as liquid boils away. Cool. Strain and add 4 ounces copperas crystals; stir until completely dissolved. Add enough water to make 4 gallons.

Add damp wool yarn (1 pound, dry) and simmer 30 minutes. Dissolve 4 tablespoons cream of tartar and ½ cup Glauber's salts in 1 pint boiling water and add to dye bath. Keep yarn covered with the bath to prevent streaking and simmer 30 minutes longer. Cool. Rinse until water is clear. Dry yarn in the shade.

Color: Dark green. Re-use dye bath for lighter shades of green.

Horsetail *(Equisetum)*:

Gather 2 or more gallons of the branching stalks (scouring brushes). Put the stalks in a pot and cover them with rainwater. Layer 1 pound wool yarn alternately with the horsetail. Use amount of yarn for each layer equal to about $\frac{1}{10}$ of the weight of the wet plant layer. Simmer for 30 minutes, rinse repeatedly in warm water until water stays clear, and dry in the shade.

The horsetail-dyed yarn can be re-dyed with other plants. Colors seem to improve if the yarn is first dyed with horsetail.

With alum mordant: Greenish yellow.

With copperas crystals (iron or ferrous sulfate) mordant: Gray green.
With copper sulfate mordant: Grass green.

If an aluminum or zinc kettle is used, a green color will result if plants are heated in the dye bath for 2 hours.

Nettle *(Urtica Lyallii)*:

Gather mature plants, using gloves and scissors. Chop whole plants into small pieces, cover with water, and boil 1 hour. Strain and add water to make 4 gallons. Heat dye bath to lukewarm. Add wet, mordanted wool yarn, bring to a boil, and simmer 30 minutes. Cool, rinse in warm water until clear, and dry.

With alum mordant: Greenish yellow.

Northern Bedstraw *(Galium boreale)*:

For ½ pound wool yarn, use ½ pound roots and 1 ounce alum. Crush the roots, wrap in cheesecloth, and soak overnight in rainwater to cover. Add enough water to make 2 gallons, then boil for 1 hour, add alum, and stir until dissolved. Add wet yarn and simmer 30 minutes, or until color is satisfactory. Rinse thoroughly and dry.

No mordant: Brownish pink.
With alum mordant: Light red.
With chrome mordant: Purplish red.

For a deeper color, use ½ the quantity of roots and wool yarn, add 4 ounces vinegar or acetic acid with the yarn, and simmer gently for 1 hour.

NOTE: The leaves produce a yellow dye.

Póochki *(Heracleum lanatum)*:

References inform us that póochki makes a dye ranging in color from light brown to yellow or gold, depending on mordants. We do not have the recipe; feel free to experiment!

Rumex acetosella:

Gather the whole plant. Put chopped plants, 4 gallons lukewarm water, 1 ounce cream of tartar, and 3 ounces dissolved alum in a wooden bowl and allow to steep 2 to 3 weeks. Add 1 pound yarn and leave in dye bath for a couple of days. Stir occasionally for an even color. Hang to dry without rinsing.

Color: Light grayish pink.

NOTE: This method can also be used for willow, mountain ash, alder, birch, and yarrow.

Sourdock *(Rumex)*:

Dig the roots anytime after the plant blooms. Late autumn is OK, if you can locate the plant. The roots, which look something like sweet potatoes, can go quite deep. They can be used fresh or dried.

Chop roots into small pieces. Cover well with water, soak overnight, then boil for 2 hours in the same water. Mash with a potato masher while they are cooking. Strain, add enough water to make 4 gallons, add 1 pound yarn, and simmer 1 hour.

No mordant: Tan.
With alum mordant: Gold.
With chrome mordant: Reddish tan to deep gold.

Silverweed *(Potentilla anserina)*:

Although several references inform us that a reddish dye can be made from the roots of silverweed, we have not found a specific recipe. We would suggest using alum-mordanted wool yarn and following a process similar to that set forth for sourdock roots.

Spreading Wood Fern *(Dryopteris dilatata)*:

We understand that this plant's fibrous, dark-brown roots yield a brown dye, but we do not have specific directions.

DYES FROM TREES AND SHRUBS

Tree bark should be collected in the spring, as the sap is highest then. Gather from young trees that are free of moss. (Do not peel bark all the way around the tree — this will kill the tree.) Dry the bark. It can then be kept for several years if stored in a dry place. As almost all tree barks contain tannic acid, the addition of iron salts gives dark colors.

Alder *(Alnus crispa)*:

Gather branches and strip off outer and inner bark. (Use 2 pounds bark to 4 gallons water.) Cut bark into small strips and soak overnight in water to cover. Boil for 2 hours in the same water. Strain, add enough water to make 4 gallons, add 1 pound yarn, and simmer 1 hour.

No mordant: Brown.
With 2 ounces alum mordant (⅔ ounce cream of tartar): Shades varying from ice green to orange.
With copperas mordant: Gray brown.

Spruce *(Picea sitchensis)*:

Bark

Peel inner and outer bark of the spruce tree and follow the same process as with alder. Simmer until the desired shade is obtained.

Color: Brown.

Twigs

Chop lichen-free twigs and simmer in water to cover, using aluminum or enamel kettle, for 12 hours. Then place yarn on top of the twigs but keep it under water. Heat to 176 degrees. Simmer, stirring to get an even color. When the yarn reaches the desired shade, hang to dry without rinsing. Longer simmering will yield a darker shade.

Color: Camel-hair tan.

Blueberry *(Vaccinium)*:

Gather berries when fully ripe. Crush ½ pound berries; boil in 1 gallon water for 1 hour to extract color. Strain. Add water to make 1 gallon. Add about 1 ounce alum and boil 5 minutes longer. Add wet wool yarn (1 pound, dry) and simmer for 1 hour.

Color: Lavender to purple (fades in sun).

Elderberry *(Sambucus racemosa)*:

Boil ½ pound berries in ½ gallon water for 30 minutes to 1 hour. Add 1 tablespoon salt. Strain and add water to make ½ gallon or more. Add wet yarn (1 pound, dry) and simmer for 1 hour.

No mordant: Red.
With alum mordant: Lilac blue.
With chrome mordant: Plum to lavender.

NOTE: Elderberry leaves produce a yellow dye.

Lingonberry *(Vaccinium vitis-idaea):*

Use fresh stems and leaves gathered in summer (use 2 gallons of the plant to produce 4 gallons of dye bath). Cover with water and let stand overnight. Boil 2 hours, strain, add enough water to make 4 gallons, then add 2 tablespoons alum and 1 tablespoon cream of tartar. Add wet yarn, stir, and simmer in the bath for 1 hour or until desired shade is obtained. Rinse repeatedly in warm water until water is clear; hang to dry.

Color: Red.

Sweet Gale *(Myrica gale L.):*

Three different shades of yellow can be obtained from the same dye bath. Divide 12 ounces wool yarn into 3 skeins and mordant with 1½ ounces alum in enough water to cover. Let wool cool in the mordant solution.

Cover 10 ounces fresh sweet gale leaves with cold water and boil 1 hour. (If dried leaves are used, soak overnight first.) Strain. Add enough water to cover yarn well and heat slowly. When temperature reaches 95 degrees, add all 3 skeins of yarn; simmer 45 minutes. Remove and hang yarn.

Color: Golden yellow.

Now pour half the dye bath into another kettle and add $\frac{1}{5}$ ounce copper vitriol; stir until dissolved. Add 1 gallon water. When temperature reaches 104 degrees, drop in 1 of the 3 skeins of yellow yarn and simmer 15 minutes. Remove and hang yarn.

Color: Warm yellow brown.

To the remaining dye, add $\frac{1}{5}$ ounce copperas crystals and stir until dissolved. Add 1 gallon water. Heat to 104 degrees and drop in 1 skein of the yellow yarn. Let simmer 15 minutes. Remove and hang yarn.

Color: Cool yellow green.

Wash and rinse the three skeins of wool yarn and hang to dry in a shady place. The three colors will blend very well.

SOURCES FOR DYE MATERIALS

Gordon's Naturals
P.O. Box 506
Roseburg, Oregon 97470

Straw-Into-Gold
P.O. Box 2904
Oakland, California 94618

A CONVERSATION WITH SASHA AND JENNY

One of the rewards I received for working on this book was recording the following interview. The ladies introduced here, Jenny Chernikoff and Alexandra Smith are, to me, beautiful women, who portray their people and their way of life far better than words could ever do.

Jenny and Alexandra (or Sasha, as we call her) are both in their late seventies. They grew up together; they are cousins; and they have been friends since childhood. They have lived in Ouzinkie all their lives, and have together seen the changes come to their home on an Alaskan island. They both remember when things were different. They consented to allow me to record this conversation so they could share some information about local plant lore with the readers of this book.

I am very pleased that they have been willing to share what they know with us. But, to me, the portrait they give of themselves is of as great a value as their plant knowledge. They are real people. They touch earth. They have a grace that we may earn if we live long enough and learn the right things along the way.

It would be nice if one could hear as well as read their words, and enjoy their accents, phrasing and inflection: Sasha's voice, low and melodious, with that subtle difference that reminds me of Ouzinkie, and Jenny's, clear and distinct, with an accent that comes perhaps from her Norwegian ancestry.

We leafed through *The Alaska-Yukon Wild Flowers Guide* and the Cooperative Extension's *Wild, Edible and Poisonous Plants of Alaska* on the afternoon the tape of this conversation was made and discussed the plants together.

Much of our conversation that day is included here. However, I have omitted dialogue about many flowers shown in the Alaska-Yukon book that both ladies admired simply because the blossoms are beautiful. They told me where they found them, and some of the adventures they had in the process. We'll save that part of the conversation for another book.

A footnote: One of the things I love best about my favorite Ouzinkie people is their sense of humor. Both Sasha and Jenny have their share of that commodity. Both of them believe laughter to be a great medicine. There is some of that humor at work in this conversation between us. For example, shortly after I had transcribed our talk from the tape, I gave the draft of our conversation to Sasha to read. I glanced over at her a few minutes later and saw her shoulders shaking in silent laughter. I went over to her. "What are you laughing at, Sasha?" I asked. "This," she replied, pointing to the place in the script. Jenny had been looking at the picture of the Canadian dwarf cornel, locally called the airberry because its fruit seems filled with air when eaten. "That's a pirdunia," Jenny had said. "That's called 'pirdunia' in Russian," Sasha had agreed. Sasha explained to me that the Russian word they had used referred to the gentle art of passing gas.

Well, anyway, here we are in Jenny's neat and cozy little living room, with family photos, plants and collectibles, and the Russian Orthodox altar high in one corner. The tape recorder and I are lazily seated on the floor; Sasha and Jenny are on the sofa, the plant books between them. The tape begins. I am pleased to introduce to you my dear friends, Jenny Chernikoff and Sasha Smith.

Jenny: Here's the chocolate lily. You can eat them; cook them, the bottoms. They're most delicious. You have to get lots of them and then you cook them, and they're really good. I don't know what they're good for, but anyway, I have heard if you have some kind of problem in your stomach, that helps some. To cook them, you just boil them.

Fran: Just like rice?

Jenny: Yeah. Make sure you have that middle part.

Sasha: She means that little wild rice that's on the bottom.

Jenny: And then, don't take off the little ones, and cook it.

Sasha: You cook them like potatoes.

Jenny: You can season it any way you want to, and then — best when you put berries with it — ahh, wild raspberries — oh, that's delicious. But they claim it has something to do with your stomach. You know, some people don't digest food the right way or something. But the flower doesn't smell very good.

Fran: How about wild chives?

Jenny: That's good for cooking, you know. You eat it; cook with it. But I don't know what it will do to you. [As medicine.] As a seasoning, it's okay.

Fran: How about the pond lily? Any uses for that?

Jenny: No, we don't use that.

Sasha: This is "my darlings."

Fran: Narcissus-flowered anemone?

Sasha: Yes, we always called them "my darlings."

Jenny: Skrípkas [roseroot].

Fran: What do you call it? Skrípka?

Sasha: You know what "skrípka" is in Russian? It's a violin. When you rub the leaves between your fingers, they make a little noise.

Fran: Have you ever eaten them?

Sasha: No, I don't think they eat them. They grow in the high places.

Jenny: See, this name I'm trying to think of is similar to this [saxifrage]. What I'm trying to think of is similar to this name. The leaves are broad. If only I found one and then I'd know it. Ah . . . they call it saparell . . . sarsaparella . . . but the leaves are similar to . . . they grow in the lake . . . what do you call them again?

Sasha: Pond lilies?

Jenny: Yes, similar to that, but it's more cut this way . . . (indicates uneven edges).

Fran: And they're tall?

Jenny: No, no, they grow about that tall (indicates low plant).

Fran: Little guys?

Jenny: Yeah, but they're very scarce here. Very, very scarce. [Author's note: This is our mystery plant. We are still hoping to learn what it is. This winter, long after this interview was recorded, Julia Pestrikoff of Port Lions gave me pressed leaves and dried roots of the plant, also calling it "sarsaparella." When Jenny and Sasha saw the leaf, they both recognized it immediately. But the commonly known "sarsaparella" plant does not grow in Alaska. What have we found this time?]

Fran: What color are the flowers?

Jenny: Oh, I don't think they bloom . . . but, I know they've been used for lots and lots and lots of medicine. And same thing with that — I don't know what you call it. It grows short, under the trees.

Fran: Star of Bethlehem? [Wax flower, or shy maiden.] The one with the little face that turns over?

Jenny: Yeah. That's used for twelve sicknesses — imagine!

Fran: Can you remember what they all are?

Jenny: Well, I know two or three of them. One is for when something is wrong with your lungs, and the other one I think is for something wrong with your stomach, and I know one lady here — that was the only thing that helped. They went to the woods and got some and they dried them by the stove. And they made a tea and had her drink it, and she drank it, and, oh, she vomit blood . . . after that she got well. Every summer we used to go pick them. I used to like to smell them.

Sasha: They smell real nice, those little plants.

Jenny: I didn't use it, though. But, I don't know . . . lots of them, they claim they're for lots of sicknesses. Now, same thing with nettles. You eat the nettles, don't you?

Sasha and Fran: Yes.

Jenny: That's the best medicine when you have toothache.

Fran: Oh really? How do you use it for toothache?

Jenny: You get the roots. Wash them real good, and pound them, and use them as a poultice.

Fran: Outside your jaw?

Jenny: Yes. Or you can take pieces of root, wash them very good, and bite down on them. But you spit out the saliva at first. I know that. I heard about it and I've seen it being used.
   Same thing with the alder — the ordinary alder. You take the top off the bark, and scrape it. Take the inside of the bark, then you scrape that until you get enough, and then you brew it. Then when you have tonsillitis or sore throat, you gargle with it, and it will help you, lots. If you want to sweeten it a little bit, that's fine.

Then, the spruce I know. When the spruce first gets cones, when they get light green. You see, they're kind of red at first, and then when they're almost prime, they'll get light green. They're soft, see? That's good when you have a chest cold, and so forth. You got to brew it.
   And then, what would you call "igória"?

Sasha: Wild geranium.

Jenny: Wild geranium is good for sore throat, too.

Fran: Do you use the leaves?

Jenny: You use the root. You wash them real good and use the root.

Sasha: They say you scrape the brown part away from there a little bit.

Jenny: Well, they claim that's the best part of it.

Sasha: I suppose, as long as you wash them real good.

Jenny: No, that's not poisonous. That's good for sore throat and chest cold and so forth.
   Oh, lupine — that's "kakoríki."
   Wild rose — that's for lots of things. The rose is for lots of stuff.

Fran: What kind of stuff?

Jenny: Well, when you don't stop coughing when you have a cold, you use that. Brew it, you know.

Fran: What part?

Jenny: Petals. There's lots of vitamin C there, you know. Besides, you can eat those pods, you know.

Fran: The hips?

Jenny: The hips, yes. Whatever you call them. I seen one person — that person didn't know what to do, what to do. He was coughing, coughing, coughing. And then he comes to my aunt and he said, "Can you help me?" He tried everything — store-bought medicine and everything else. And then my old aunt, she asked me if I had any red rose hips. Well, I had those — I'd dried them. I took them over. That person was still there. She fixed it. She told him, "Keep on, keep on, keep on taking it." Next thing, that person, he thank her so much. He's still alive, that person.

Violets! You can't use them for nothing, though. I don't know about using them.

Fran: You don't think they used them here?

Jenny: No.

Fran: You can eat them, you know. They have lots of vitamin C.

Jenny: You mean, violets? Oh, no wonder there's no more violets in Ouzinkie, then.

Fran: How about fireweed?

Jenny: You can make "chai" (tea) out of it.

Fran: Do you use the leaves for chai?

Jenny: Yes, you dry them.

Sasha: Water hemlock — that's poisonous. That's no good.

Jenny: Cow parsnip — *that's* not poisonous — cows eat them.

Fran: There's the Canadian dwarf dogwood.

Jenny: Pirdunias. That's a pirdunia.

Fran: Is that the one you call airberry?

Sasha: Yes, that's their blossom. That's called "pirdunia" in Russian.

Jenny: [The plant I'm trying to remember is] similar to pyrola but the leaves are oval — the leaves are cuplike. I can't remember them — sarsaparella — I know, I went out one time with old lady Irene, and we picked some. Big Lagoon — we walked all the way to Big Lagoon [on the back side of Spruce Island], and at Big Lagoon at the top of the hill, there's a big lake there, great, big lake — dark one. I remember around there somewhere we got a bunch. And we send them to Kodiak. They brew that into cough syrup.

And then, that mogúlnik (Labrador tea). That is a sure cure for chest and so forth.

There was a priest's wife, and we used to pick mogúlnik by sacks. She had T.B. — it's pitchy, you know. It's pitchy, and she took it, and there was nothing wrong with her when she died.

Yes, there's [the picture of] star of Bethlehem — good for lots and lots of sicknesses — but you dry them and brew them.

Fran: How about Jacob's ladder?

Jenny: Oh, yes, they grow here, too.

Sasha: Yes, there's lots of them here.

Jenny: One time she and I found a white one.

Sasha: And then we lost it next.

Jenny: Oh, we got all excited, and took roots and all up. And then (laughs), we lost them. We were out in the swamp, here, and next thing we know we'd lost our plants. We tried to go and look for them but we could never find them.

Sasha: No, we could never find them. Boy, we wanted them. They were nice and white.

Jenny: Yes, well, they would look nice with the blue ones.

Sasha: Where did you find them, now?

Jenny: On that side of the mountain — on the mountainside. Oh, look, here's some more. Kushelkoks — they're called kushelkoks.

Sasha: There's all kinds of little flowers growing up there on the mountain.

Fran: We'll have to go up on the mountain this summer, and find some plants we haven't seen before.

Sasha: I would like to.

Jenny: Are you going to take me along?

Fran: Well, of course!

Jenny: What are you going to do with the bear?

Sasha: The bear wouldn't come.

Fran: We'll scare him. We'll make so much noise he won't bother us. We'll take Rosemary's dogs along, too.

Jenny: They say there's two.

Sasha: What's that?

Jenny: Two bear! Two of them was at Mahoonah. (A big lake at the center of Spruce Island.) If they're at Mahoonah, they're wandering close by here again. Those things wander fast!

Sasha: Yes, Spruce Island is small.

Jenny: I told you the other time something was wrong, and I didn't want to say nothing.

Sasha: I know, you didn't have to tell me.

Jenny: Did you know?

Sasha: Yes, I didn't say nothing to you either because I know you would get scared.

Fran: You thought that there was a bear over there?

Sasha: Yes, something had been scratching up the ground.

Jenny: Coastal paintbrush — I can't think — they used them for something. They were almost like they were already dried. But I can't think what they were for.
   Poque — that's a bear food. Bear eat that.

Fran: We'll pick them all so the bears will leave.
   That's goldenrod. Do they use goldenrod for anything here?

Jenny: "Zhólti golóvnik" is the Russian name. "Zhólti golóvnik." "Zhólti" means yellow.

Fran: Seabeach senecio — Can you remember anything about that? I've heard it was used for medicine.

Jenny: We were told not to touch them when we were small.

Sasha: Some kind of medicine, then.

Jenny: I really don't know. We always wanted to pick the flowers, but they told us not to touch them, and they made us wash our hands if we did. I still don't touch them. I just don't do it. Once in awhile maybe I step on them.

Fran: Would you pronounce the Russian name for goldenrod again?

Sasha: "Zhólti golóvnik." I think they used to use it for a whip when they have the hot bath. I think they use it for some kind of an enema.

Jenny: "Kutágarnik" (one of the tall wild parsleys). Not edible. That's different from petrúshki (beach lovage). In fact, I have some right now in my yard.

Sasha: That's that poisonous kind that I tell you not to eat.

Jenny: No, it's not poisonous because cows eat it. The reason why I have some in my yard is because when they got tall I used to cut it and keep it; feed it to the cows in the winter time.

Fran: That's spring beauty.

Sasha: They used to tell us not to touch it or it would rain.

Fran: Is that the one you call "rain flowers"?

Sasha: Yes.

Fran: Oh, really?

Jenny: "Lóstochki," in Russian.

Fran: What do you call fireweed in Russian?

Jenny: In Aleut it's "chílchkoch." "Chílchkoch." What do they call them in Russian, then? (Aleut spelling: "Cillqaq.")

Sasha: I really don't know what they call them in Russian.

Jenny: The leaves on "sarsaparella" are similar to coltsfoot. I think they have some kind of little blossom. I'm not sure, though.
    Nettles — I know they're very good for toothache. Put the nettles in a heated rag when you use them for a poultice. I used it before — I helped with it.

Sasha: Blackberries (crowberries) — "shíksha." "Shikshónik" is the crowberry bush.

Fran: Have you ever seen wild strawberries here?

Jenny: There was some other side here, but I don't know how they got here. Other side where the dump is. Before the hill wasn't that way. [Before the tidal wave, when the hill slid into the sea.]

Sasha: I remember them. I thought they were those tame ones. They were real small.

Jenny: I couldn't figure out where they came from. Boy, the berries would get real big. But they all slid down now, into the ocean. I don't know where they came from.
    Nagoonberry —

Sasha: "Puyurniqs," they call them in Aleut. Cloudberry is "maróshka." Salmonberry is "malína." (In Russian.)

Fran: Aren't there some things you can use salmonberries for?

Sasha: A friend told me you can use them for some kind of sore that won't heal. You take the old leaves out from under the bushes. She's told me many times.

Jenny: Long ago, you know, there were no doctors, and people would rely on what grew, you know. And they would practice, and see how it would work, and so forth like that. And then, but the Aleuts had a different way of doing things, not like the way we were brought up.

Fran: How about "romáshka?" (Pineapple weed.)

Sasha: They're a mild laxative, I guess.

Jenny: If you need to clean your insides, you know, like after childbirth. Long ago they used to use that to cleanse the mother out. After childbirth, the mother is constipated, so they brew some of that stuff to clean her out. That's a nice thing to know about. Once in awhile I still have that. And you can use milk in it, or sugar, or whatever.

Sasha: I like that kind of tea. Once in awhile, I drink it, too. I made some last summer.

Jenny: The best part is the lower part. You cut off the root. When we used to pick them, we were allowed to cut off so much and left the root in the ground, so next year they'd come up again. Or we'd leave one with lots of blossoms, and after that, it would self-seed, see? It's getting pretty scarce here now. I don't see it much here anymore. I know back in the garden there used to be lots of them and I used to save them; you know, make them get a little bit taller. But I don't see none anymore.

Sasha: There was lots of them in my garden last summer. I made a bunch for her and then I made some for myself, too.

Jenny: That's really good stuff. I really can't think of the one I want, that's been used for many years here . . . I know, "igória," that's the main one for sore throat.

Fran: What about nizamýnik? (Devil's club.)

Jenny: I really don't know.

Sasha: I think nizamýniks are good for colds, because Leo [her grandson] was telling me their uncle used to let him drink that juice after he makes it when they are small. They take the roots and then boiled them, I guess, and then he put it in a jug, I guess, and then when they had a cold he'd let them drink it kinda warm, that kind, and it loosens the chest cold. But you only use a little bit, because it can make you kind of drunklike.
  "Polín" — there's "polín" (caribou leaves). They smell like spearmint.

Jenny: "Polín" is used if you're crippled up with arthritis or something, and you can't walk. Then they use that, heat it up and put it on wherever you're hurting.
  If I see it, I know it. I think there's some at Mahoonah. If I walk around in the woods, I'd know some more.

Fran: We'll do that together, when things are all up.

Sasha: June or July sometime would be a good time, when they're all growing . . .

GLOSSARY

Algae (singular alga) — Any of a group of lower plants having chlorophyll but no vascular system, such as stem or leaf veins, for carrying essential fluids. Seaweeds and related freshwater plants are examples of algae.

Alternate — Refers to a way leaves grow on the stem of a plant. Alternate leaves are arranged singly at different points along both sides of the stem.

Annual — Refers to the life cycle of a plant. An annual completes the cycle from seed to death in one year or season.

Antiseptic — An agent or substance for destroying or slowing down germs that cause decay and infection.

Archipelago — A group of islands.

Astringent — An agent or substance that shrinks body tissue, reducing secretions or discharges.

Biennial — Refers to the life cycle of a plant. A biennial completes the cycle from seed to death in two years or seasons.

Bract — A modified leaf, usually smaller than the true leaves. It forms either on the flower stalk or as a part of the flower head. Bracts are sometimes mistaken for flowers.

Cambium — A thin cellular layer of tissue, under the hard outer bark of trees and shrubs, from which new tissues develop.

Carbohydrate — Any of various compounds made up of carbon, hydrogen, and oxygen. Sugars and starches are such compounds.

Catkin — A spikelike flower cluster (as of a willow) bearing petal-less, unisexual flowers and having bracts, or modified leaves, on or at the base of the flower stalk.

Caulk — To make tight against leakage by a sealing substance; to make the seams of a boat watertight by filling with waterproofing materials.

Chlorophyll — The green material in plants. When exposed to sunlight, chlorophyll forms carbohydrates in a process called photosynthesis.

Coagulate — When a portion of liquid (for example, blood) thickens and sticks together, it clots, or coagulates.

Companion plants — Plants which grow together or which are planted together because their physical demands for growth complement each other.

Decoction — To make a decoction, place the plant parts to be used in water and boil them together. This method is used to extract mineral salts or bitter substances from plants, or to remove active ingredients from hard materials such as roots, bark and seeds.

Diuretic — An agent that increases the flow of urine.

Dropsy — An abnormal accumulation of fluid in the body.

Family — A group of related plants having common characteristics. A family of plants may include many plant genera (see genus).
Frond — The leaf of a fern.
Fungus (plural fungi) — Any of a large group of lower plants that do not contain chlorophyll. This group includes molds, mildews, mushrooms, and bacteria.
Generic — Of or relating to a genus.
Genus (plural genera) — A grouping of plants of closely related species. Families of plants are divided into the more specialized groups of genus and species. The division into genus and species is called the "binomial system," and uses a pair of Latin words for each plant name. The first part of this botanical name is the genus of the plant and is always capitalized. The second word is the species name and is usually not capitalized. The species is the fundamental group that can be consistently identified as being different from other plant groups.
Habitat — The place or kind of place where a plant naturally grows.
Hardy — Plants that are able to withstand adverse weather and freezing temperatures.
Hemorrhage — A large discharge of blood from the blood vessels.
Hemorrhoids — A swollen mass of dilated veins situated at or just within the anus.
Indentations — Notches or deep recesses, as in leaf edges.
Infusion — An infusion is a beverage made like tea, by combining boiling water with plant parts (usually the green parts or the flowers). The plant parts are then steeped — left to soak in the hot water — to remove their active ingredients. This method, with its relatively short exposure to heat, minimizes the loss of vitamins or other ingredients that are destroyed by cooking.
Lobe — A part or division of a leaf that is curved or rounded.
Mordant — A substance used in dyeing for the purposes of fixing the color and making it permanent.
Node — The place on the plant stem where a leaf grows or can grow.
Opposite — Leaves growing two to a node on opposite sides of the stem.
Perennial — A plant that lives through three or more seasons.
Photosynthesis — The process by which chlorophyll-producing plants form carbohydrates when exposed to sunlight.
Pollinate — To carry pollen to the female part of a plant to fertilize the seed.
Poultice — A soft, usually heated mass made from medicinally useful plants, then spread on or wrapped in cloth and applied to a sore or injury.

Regenerate — To form or create again. Plants regenerate themselves in various ways.

Respiratory — Referring to the act or process of breathing. Diseases of the respiratory system would be those which interfere with this breathing process.

Rhizome — An underground portion of a plant stem, having shoots on top of it and roots beneath it. Though it is underground, it is different from a root in that it has buds, nodes, and scaly leaves.

Rosette — A circular or spiral arrangement of leaves growing from a center or crown. Many plants have leaves arranged in a "basal rosette," or a circular arrangement of leaves at the base of the plant.

Scurvy — A disease caused by a deficiency of vitamin C. Symptoms include spongy and bleeding gums and loose teeth.

Silica — A mineral composed of silicon and oxygen. Silica is found in some plants, and can be useful in treating certain physical conditions.

Species — The basic unit of scientific classification of plants. The species name is often descriptive of some distinctive quality of the plant. There can be numerous species in a genus of plants, and many genera of plants in a family.

Steep — Allowing herbs to stand in hot water to extract their active ingredients. Herbs should be steeped in an enamel, glass, or porcelain pot, not a metal one. A covered container is best.

Succulent — A plant having fleshy tissues that conserve moisture.

Symbiotic — A type of association or union of two differing organisms which live together for mutually beneficial purposes.

Textile — Any kind of cloth, especially woven or knit cloth.

Tonic — An agent or substance that strengthens or invigorates organs or the entire body.

Toxic — Poisonous.

Vascular — In plants, refers to the system of channels for carrying life-giving fluids. These channels include plant stems and leaf veins. In the human body, the vascular system carries blood through the body.

Vitamin — Any of various organic substances that are essential, in tiny amounts, to most animals and some plants and are mostly obtained from foods.

Whorl — A circular arrangement of three or more leaves or flowers at the same point or level on the stem of the plant.

BIBLIOGRAPHY

Alvin, Kenneth L. *The Observer's Book of Lichens.* London: Frederick Warne Ltd., 1977.

Angier, Bradford. *Feasting Free on Wild Edibles.* Harrisburg, Pennsylvania: Stackpole Books, 1966.

Angier, Bradford, *Field Guide to Edible Wild Plants.* Harrisburg, Pennsylvania: Stackpole Books, 1974.

Ball, Georgiana. "Medicinal Leaves of the Tahltans." *ALASKA®*, vol. XLIX, no. 4, April 1983.

Benoliel, Doug. *Northwest Foraging.* Seattle: Signpost Books, 1974.

Bricklin, Mark. *Natural Healing.* Emmaus, Pennsylvania: Rodale Press, 1976.

Chase, Cora G. *The Weedeater's Cook Book.* Seattle: Shelton Publishing Co., 1978.

Christiansen, Clyda. "Health Remedies." *Kodiak Area Native Association Newsletter*, 1982.

Clark, Lewis J. *Wild Flowers of Marsh and Waterway in the Pacific Northwest.* Sidney, British Columbia: Gray's Publishing Ltd., 1974.

Clark, Lewis J. *Wild Flowers of the Sea Coast in the Pacific Northwest.* Sidney, British Columbia: Gray's Publishing Ltd., 1974.

Cobban, Gerry. "Island Edibles and Elixirs." *Memoirs of a Galley Slave*, Fishermen's Wives Auxiliary. Kodiak: Page Photo, 1981.

Coon, Nelson. *Using Plants for Healing.* Emmaus, Pennsylvania: Rodale Press, 1963.

Cooperative Extension Service. *Wild Berry Recipes.* Fairbanks: The University of Alaska, 1973.

Cooperative Extension Service. *Wild, Edible and Poisonous Plants of Alaska.* Fairbanks: The University of Alaska, 1966.

Domico, Terry. *Wild Harvest*. Seattle: Hancock House Publishing Inc., 1979.

Editors of *ALASKA®* magazine. *The Alaska-Yukon Wild Flowers Guide*. Anchorage: Alaska Northwest Publishing Company, 1974.

Gibbons, Euell. *Stalking the Healthful Herbs*. New York: David McKay Company Inc., 1966.

Grieve, Mrs. M. *A Modern Herbal*, vols. I and II. New York: Dover Publications Inc., 1971.

Hall, Alan. *The Wild Food Trail Guide*. New York: Holt, Rinehart and Winston, 1973.

Hall, Nancy and Walter. *The Wild Palate*. Emmaus, Pennsylvania: Rodale Press, 1980.

Hulten, Eric. *Flora of Alaska & Neighboring Territories: A Manual of the Vascular Plants*. Stanford, California: Stanford University Press, 1968.

Hylton, William H., ed. *The Rodale Herb Book*. Emmaus, Pennsylvania: Rodale Press Book Division, 1974.

Kari, Priscilla Russell. *Dena'ina K'et'una, Tanaina Plantlore*. Anchorage: University of Alaska, Adult Literacy Laboratory, 1977.

Keats, Della. "To Heal the People." A reprint in part from *Northwest Arctic Nuna*, Fall Edition. Kotzebue, Alaska: a Maniilaq Association Publication, 1982.

Las Aranas Spinners and Weavers Guild. *Dyeing with Natural Materials*. Albuquerque: Las Aranas, 1973.

Leer, Jeff. *A Conversational Dictionary of Kodiak Alutiiq*. Fairbanks, Alaska: Alaska Native Language Center, University of Alaska, 1978.

Lesch, Alma. *Vegetable Dyeing*. New York: Watson-Gupstill Publications, 1970.

Lust, John. *The Herb Book*. New York: Bantam Books, 1974.

McMullen, Elenore. "Home Remedies We Have Used." *Fireweed Cillqaq*. Port Graham High School Publication, vol. 2, Soldotna, Alaska: Kenai Peninsula School District, 1981.

Meyer, Joseph E. *The Herbalist*. 1918. Reprint. Glenwood, Illinois: Meyer Books, 1975.

Oberg, Kalervo. *The Social Economy of the Tlingit Indians*. Seattle and London: University of Washington Press, 1973.

Preston, Eudora M. "Medicine Women." *ALASKA SPORTSMAN®*, November 1961.

Prevention. Emmaus, Pennsylvania.

Ray, Glen. *Root, Stem and Leaf: Wild Vegetables of Southeast Alaska.* Juneau: Southeast Regional Resource Center, 1982.

Robinson, Peggy. *Profiles of Northwest Plants.* Portland, Oregon: Far West Book Service, 1978.

Rose, Jeanne. *Herbs and Things.* New York: Grosset and Dunlap, 1972.

Schetky, Ethel Jane McD. and Carol H. Woodward. *Dye Plants and Dyeing.* Special printing of "Plants and Gardens," vol. 20, no. 3, Brooklyn, New York: Brooklyn Botanic Garden, 1964.

Spellenberg, Richard. *The Audobon Society Field Guide to North American Wildflowers.* New York: Alfred A. Knopf, 1979.

Stuart, Malcolm, ed. *Herbs and Herbalism.* New York: Van Nostrand Reinhold Company, 1979.

The Herb Quarterly. Newfane, Vermont, 1979-1981.

Viereck, Leslie A. and Elbert L. Little, Jr. *Guide to Alaska Trees.* Forest Service, United States Department of Agriculture, 1974.

Wren, R.C. *Potter's New Cyclopaedia of Medicinal Herbs.* New York: Harper Colophon Books, 1972.

INDEX

Aches, 56
Achillea borealis, 28-29
Aconitum delphinifolium, 135-136
Actaea rubra, 134
Adiantum pedatum, 84
Afognak Island, 112
Agar, 87
Air purification, 120
Airberry, 91-92
Alagnaq, 115-117
Alaska miner's lettuce, 71
Alaska spring beauty, 2, 71-72
   hints on cooking, 144-145
   recipes using, 65
Alder, 128-129, 175
   dye from, 170
Alder trunks, ferns on, 83
Aleut
   doctors, 6-7
   medicine, 179
   Pilaf, 148
   plant names, 2-4
   plant uses, 141
Aleutian Islands, plants in, 136
Algae, definition of, 181
Allium schoenoprasum, 42-43
Alnus crispa, 128-129
   dye from, 170
Alternate, definition of, 181
Alum mordant, 165
Amaryaq, 106
American cranebill, 33
American red raspberry, 112
Anemia, 38
Anemone, 29
Anemone, narcissus-flowered, 136
Anemone narcissiflora, 136
Angelica genuflexa, 140
Angelica lucida, 140
Animals, food for, 60
Annual, definition of, 181
Antibiotics, 120
Antiseptic, 131, 181
Appetite, increasing, 29, 73
Aramaaskaag, 26-27
Archipelago, definition of, 181
Arctic dock, 10-14
Arctic explorers, 64

Arctostaphylos alpina, 103
Arctostaphylos uva-ursi, 103
Aromáshka, 26-27
Artemisia, 16
Artemisia tilesii, 16-18
Arthritis, 18, 32, 38, 40, 56, 119
Asparagus, 32
Aspirin, pain reliever in, 132
Asthma, 29, 75
Astringent, 12, 26, 33, 60, 103, 115, 121, 128, 132
   definition of, 181

Babies, 27, 33, 40, 112, 130
Baby's breath, Alaskan, 45
Bad breath, 73
Baidarka Soup, 153
Baked Deer Spareribs, 149
Baked Halibut with Mushrooms, 157
Baldness, 29
Band-Aids, temporary, 130
Baneberry, 29, 134
Banjo Beach Omelette, 20
Barbecued Fish Fillets, 161
Bark, edible, 146
Barrel hoops, 121
Basic Cream Soup Recipe, 144
Basic Puree Recipe, 144
Baskets, 131, 132
Bass, French Fried Sea, 156
Batter, Fish in French, 158
Bay leaves, tea from, 162
Beach greens, 63-65
Beach lovage, 52-54, 148
Beach peas, 59
   hints on cooking, 142, 144, 145
Beach strawberry, 108-110
Bean Boil, Deer, 149
Bearberry, 103
Bears, 68
Beavers, 80
Bedstraw, northern, 45
   dye from, 168
   hints on cooking, 145, 146
Beer
   flavoring for, 127
   spruce, 130
Benzoic acid, 97

Berries, 87-117
   list of, 88-89
Berry
   Fritters, Wild, 96
   Sauce, Wild Fruit or, 114
Bibliography, 184-186
Biennial, definition of, 181
Bilberry, 93
Bile, 19, 28
Birch Family, 128-129
Bird nests, 85
Bites, 66, 78
Bitter plants, cooking, 145
Blackberry, 90-91, 179
Bladder infection, 45
Blanched, plants to be, 145
Blankets, 31
Bleeding, 15
Bleeding, internal, 26, 49, 84
Blender Mint Sauce, 47
Blisters, 11, 56
Blood
   circulation, 17
   clotting and coagulation, 40, 45, 49, 66
   poisoning, 45
Blossoms, edible, 99
Blotches on skin, 56
Blueberry, 93-97
   bog, 93
   dye from, 170
   early, 94
   forest, 94
   mixing with crowberries, 90
Blueberry
   Jam, 96
   Jelly, 95
   Pie, 94
   Pudding, 95
Boiled Sea Lion, 148
Boiled Wild Greens, 14
Boils, 119, 127
Boozínik, 104-105
Borage, 40
Bracken Family, 82
Bracken fern, 82
Bract, definition of, 181
Braised Dandelion Greens, 22

188 INDEX

Bread, 60
Breathing, difficulty in, 84
British Columbia, 16, 17, 129
Broad-leaved plantain, 66-67,
    dye from, 167
    hints on cooking, 144, 145, 146
Broth, Scallop, 154
Bruises, 62, 78
Brusníka, 97
Buckwheat Family, 9-14
Bunchberry, 91-92
Bunions, 66, 81
Burgers, Deer, 148
Burns, 11, 78, 116, 130
Buskin River, 17
Buttercup, 29, 135

Cake (Rose-topped Cheesecake), 125
Cake, Cranberry, 101
Calcium, 11, 19, 34, 104, 109
Caltha palustris asarifolia, 29-30
Camas, death, 138
Cambium, definition of, 181
Camping aids, 40
Canadian dwarf cornel, 91-92
Cancer, 128
Candied Silverweed Roots, 75
Candles, scented wax for, 127
Candy, Spearmint, 47
Canned Salmon with Garlic, 160
Canned wild plants, 85, 93
Carbohydrate, definition of, 181
Carbohydrates, 74, 80
Caribou leaves, 16-18, 180
Casserole, 37, 41, 43, 116, 160
    Dandelion Cheese, 23
    Fish, 155
    Nettle, 51
Cat's Crab Dip, 153
Cataracts, 90
Categories of plants, 1
Catkin, definition of, 181
Caulk, definition of, 181
Caulk for boats, 131
Celery substitutes, 53, 57
Celery, wild, 54-58
Cereals, plants for, 35, 80, 146
Chai, 38
Chamomile, 40
    tea from, 162
Chamomile, wild, 26-27
Chapped hands, 29
Cheese Casserole, Dandelion, 23
Cheesecake, Rose-topped, 125
Chenopodium, 34-37

Cherníka, 93-97
Chest ailments, 38
Chewing gum, 47, 131
Chicken Waldorf Salad, Rose-, 125
Chickweed, 62-63, 144, 145
Chiffon Pie, Lingonberry, 99
Childbirth, 27, 33, 49, 112
Children, medicine for, 46
Chives, wild, 42-43, 173
    hints on cooking, 144, 145
    recipes using, 36, 53
Chlorophyll
    commercial, 49
    definition of, 181
Chocolate lily, 41, 173
Chowder, Clam, 154
Chowder with Petrúshki, 157
Christmas greens, 14-15
Chrome mordant, 166
Chughelenuk, 115
Cicuta douglasii, 139-140
Cicuta mackenzieana, 139-140
Cigarettes, 47
Cillqaq, 31-33
Cinnamon, tea from, 162
Cinquefoil, silver, 74-76
Ciquq, 131
Cladonia dye, 167
Clam
    Chowder, 154
    Fritters, 154
    salad, 154
    Soup, 13
Clay's Crab Balls, 152
Claytonia acutifolia, 2, 71
Claytonia sibirica, 2, 71-72
Claytonia tuberosa, 2, 71
Cloth, 49
Cloth of gold, 15
Clothing, 31
Cloudberry, 113
Clover roots, 145
Clover, 60-62
    hints on cooking, 145, 146
    recipes using, 32
Clover-Bright Salad, 61
Club moss, 14-15
Coagulate, definition of, 181
Cockle Clam Salad, 154
Codfish Bacon Bake, 155
Coffee substitutes, 19, 22, 45, 146
Colds, *throughout text*
Colic, 92
Collecting plants, 5
Columbine, 29

Comfrey, 40
    tea from, 162
Commercial crops, 47
Companion plants, 27
    definition of, 181
Composite Family, 16-29
Conioselinum chinense, 140
Conserve, Wild Strawberry-
    Pineapple, 110
Conservation, traditional, 5
Constipation, 17, 62
Conversation with Sasha and Jenny,
    172-181
Cooking Instructions for Seal, 147
Cooking plants, 6
Cooking wild edibles, hints on,
    143-146
Copperas crystals, 165, 166
Cordials, 98, 111, 116
Corn lily, 138-139
Cornel, Canadian dwarf, 91-92
Cornine, 92
Cornus canadensis, 91-92
Corns, 81
Coronary thrombosis, 60
Cosmetics, 47
Cottonwood, 131
Coughs, 60, 66, 83, 84, 119, 122
Cow Parsnips, Sweet, 57
Cow parsnip, 54-58
Cowslip, 29-30
Crab, Spaghetti a la King, 152
Crab
    Balls, Clay's, 152
    Dip, Cat's, 153
    Meat Rolls, 151
    sandwich, 152
    Soufflé, King, 153
Crampweed, 74-76
Cranberry, 97-102
    bog, 97, 98
    highbush, 106
    lowbush, 97, 98
    swamp, 98
Cranberry
    -Banana Jam, 100
    Cake, 101
    Flip, 99
    Gelatin, Highbush, 108
    Jelly, 100
    Jelly, Highberry, 107
    Muffins, 102
    Relish, Crimson, 102
Cranebill, American, 33
Cream Soup Recipe, Basic, 144

INDEX 189

Creamed soups, plants for, 146
Creamed Sourdock, 13
Crowberry, 90-91, 179
Crowberry Pie, 91
Crowfoot Family, 29-30, 134-136
Cuawak, 93-97
Cucumber Salad, Wild, 44
Cucumber, wild, 43-44
    hints on cooking, 144, 146
    recipes using, 65, 72
Cukilanarpak, 118-120
Curly dock, 10-14
Cuts, *throughout text*

D.T.'s, 27
Damping-off, preventing, 27
Dandelion, 18-25, 62
    hints on cooking, 144, 145, 146
    recipes using, 61, 69, 72
Dandelion
    Blossom Pie, 24
    Cheese Casserole, 23
    Coffee Substitute, 22
    Recipe, Elizabeth Insley's, 23
    Wine, 24
Dandruff treatment, 49, 118, 132
Danny's Salmon Loaf, 159
Danny's Spicy Steamed Mussels, 155
Death camas, 138
Decoction, definition of, 181
Deer
    Bean Boil, 149
    Burgers, 148
    dishes, 147
    heart, 148
    Roast, 149
    Spareribs, Baked, 149
    Swiss Steaks, 148
Deer Fern Family, 83
Delirium tremens, 27
Deodorant, 27
Dessert, Eskimo, 64
Desserts, 78, 93, 109, 113, 116
Devil's club, 118-120, 180
Deviled Eggs with Petrúshki, 54
Diabetes, 49, 119
Diarrhea, 26, 33, 73, 75, 83, 90, 112, 128
Digestion, 46, 75
Dip, Cat's Crab, 153
Diuretic, definition of, 181
Dizziness, 38
Dock, 10-14
Dogwood, dwarf, 91-92
Dogwood Family, 91-92

Dressing, Summer Seed, 36
Dressing for poultry, 47
Dried wild plants, 85
Drinks, 99, 107, 111, 116, 124,
    see also Cordials, Tea, Wine
Dropsy, 15
Dropsy, definition of, 181
Druids, 15
Drying plants, 28
Dryopteris dilatata, 85-86
    dye from, 169
Duck
    liver, 148
    Soup, 150
Dusting powder, 15
Dwarf dogwood, 91-92
Dye
    equipment, 165
    materials, sources for, 171
    plants, preparation of, 164
    procedure, 166
Dyes from Wild Plants, 164-171
    lichens, 166-167
    small plants, 167-169
    trees and shrubs, 170-171

Easy Fiddlehead Cheese Bake, 86
Easy Salmon Bake, 161
Echinopanax horridum, 118-120
Eggs with Petrúshki, Deviled, 54
Elder, red-berried, 104-105
Elderberry, 104-105
    dye, 170
    Jelly, 105
    Wine, 105
Elizabeth Insley's Dandelion
    Recipe, 23
Elk Mulligan, 150
Empetrum nigrum, 90-91
Epilobium angustifolium, 31-33
Equisetum, 39-41
    dye, 168
Eskimo
    food, 10, 12, 64, 74, 132
    plant uses, 136
    potato, 7
Evening Primrose Family, 31-33
Eyewash, 15, 27, 29, 33, 40, 78
Eyelids, inflamed, 119
Eyes, 17, 90, 122, 130

False hellebore, 138-139
Family, definition of, 182
Ferns, 82-86, 169
Ferrous sulfate, 165

Ferrous sulfate mordants, 166
Fever, 18, 75, 92, 104, 118, 119, 122, 128, 132
Fiálka, 78-79
Fiddlehead Cheese Bake, Easy, 86
Fiddlehead ferns, 48
Fiddleheads, 82, 85, 86, 144, 145
Fir club moss, 14-15
Fire-starter, 31, 131
Fireweed
    Honey, 32
    shoots, 145
    Shoots, Marinated, 32
    stems and leaves, 144
    tips, 69
Fireweed, 31-33, 176, 179
    recipes using, 20, 65
Firewood, 129
Fireworks, 15
Fish,
    removing smell of, 27
    seasoning for, 53
    smoking, 129, 131
Fish
    Casserole, 155
    Chowder with Petrúshki, 157
    eggs, 31, 41
    Fillets, Barbecued, 161
    fillets, recipe for, 156
    Head Soup, 158
    in French Batter, 158
    soup, 50
    Soup, Mountain Sorrel and, 10
Fisharoni Suprise, 160
Flag, wild, 137
Flea collar, 127
Flour, 35, 42, 80, 121, 132
Flour, making from seeds, 144, 146
Flu, 18, 104
Forager's Fandango, 37
Forest blueberry, 94
Formic acid, 48
Fragaria chiloensis, 108-110
Franny's Favorite Spruce Island
    Weed Salad, 65
Freckles, 75
French Fried Sea Bass, 156
Fritillaria camschatcensis, 41
Fritters, 19
    Clam, 154
    Wild Berry, 96
    Wildwood, 69
Frond, definition of, 182
Frozen food, 93, 111, 112, 113, 122
Fruit or Berry Sauce, Wild, 114

INDEX

Fungi, 11
Fungus, definition of, 182

Galium boreale, 45
   dye from, 168
Gallbladder problems, 29
Game, marinade for, 39
Game recipes, 147-150
Gargle, 17, 40, 73, 75, 112, 120, 131
Garnishes, 78
Gas, 26, 29, 46, 66, 128
Gelatin, Highbush Cranberry, 108
Generic, definition of, 182
Genus, definition of, 182
Geranium, wild, 33, 175
Geranium erianthum, 33
Geranium Family, 33
Geum, 72-73
Geum macrophyllum (Avens), 72
Giant fireweed, 31
Ginger, in tea, 162
Ginseng Family, 118-120
Glauber's salts, 165
Glossary, 181-183
Goldenrod, 4, 25-26, 178
   dye from, 167
   tea from, 162
Goosefoot, 34-37
Goosetongue, 68-69
   hints on cooking, 144, 145, 146, 147
Goosetongue Royale, 68
Gout, 19
Great willow herb, 31-33
Green bean substitute, 68
Green Noodles from Mars, 21
Greens, beach, 63-65
Gum, chewing, 47, 131
Gumboot
   Rice in Gravy, 154
   Soup, 153
Gums,
   inflamed, 40
   sore, 75

Habitat, definition of, 182
Haida remedies, 118
Hair rinse, 27, 29, 49, 112-113
Hair tonic, 49
Hair treatment, 118
Halibut, Poached, 156
Halibut
   dishes, 151
   Supreme, Nellie's, 157
   with Mushrooms, Baked, 157

Hangover, 12, 29, 38
Hardy, definition of, 182
Harvesting plants, 5
Hats, 131
Headaches, 15, 18, 49, 97, 132
Heart, deer, 148
Heart trouble, 18, 33, 60, 130
Heartburn, 38, 46
Heath Family, 38-39, 93-103
Hedysarum mackenzii, 59, 142
Hellebore, false, 138-139
Helleborin, 30
Hemorrhage, definition of, 182
Hemorrhoids, 66, 121
   definition of, 182
Heracleum lanatum, 54-58
   dye from, 169
Herbicide, 115
Herbs, 9-81
Herpes, 15
Highbush cranberry, 106-108
Highbush Cranberry Gelatin, 108
Highbush Cranberry Jelly, 107
Highbush Punch, Salmonberry-, 107
Hikers, food for, 83
Hoarseness, 66, 84
Honckenya peploides, 63-65
Honey, Fireweed, 32
Honey, substituting for sugar, 87
Honeysuckle Family, 104-108
Horsetail, 39-41, 69
   dye, 168
   shoots, 145
Huckleberry, black, 94
Hudson Bay tea, 38-39
Hunters, seal, 49

Igória, 33, 175
Indentations, definition of, 182
Indian
   foods, 57, 82
   plant uses, 103
   remedies, 92
Indian rice, 41-42, 146
Indian Rice a la Russ, 42
Indigestion, *see* Stomach troubles
Infected cuts, 17
Infections, 29, 45, 104, 115, 119
Infusion, definition of, 182
Insect bites, 26, 45
Insect repellent, 27, 29, 127
Internal bleeding, 17, 26, 29, 40, 49, 84
Inupiaq, 16
Iris, 137

Iris Family, 137
Iris setosa, 137
Iron, 11, 48, 109
Iron sulfate, 165
Iron sulfate mordants, 166

Jacob's ladder, 70, 177
Jam, 93, 109, 116, 121, 122
   Blueberry, 96
   Cranberry-Banana, 100
   Mossberry, 114
   Rose Hip, 123
   Salmonberry, 117
Japan, plants eaten in, 82
Jellies, *throughout text*
Jelly,
   Blueberry, 95
   Cranberry, 100
   Elderberry, 105
   Highberry Cranberry, 107
   Nagoonberry, 111
   Rose Petal, 124
   Violet, 79
Joint grass, 39-41
Joints, stiff, 19

Kakoríka, 141, 175
Kalína, 106
Kamchatka lily, 41
Karluk, 139
Kenai people, 120
Kenegtaq, 97
Ketchup, 122
Kidney problems, 17, 26, 52, 90, 103, 130
Kidney stones, 45
Kilitáyka, 81
King's crown, 77-78
Kinnikinnik, 103
Kipráy, 31-33
Kislítsa, 10
Kléver, 60-62
Knee problems, 15
Kodiak, vi
Kodiak Historical Society, 2
Kodiak area, 2
Kostianíka, 113
Krapéva, 48-51
Kraut, 64
Kushelkok, 70, 177
Kutágarnik, 178

Laaqaq, 41
Labrador tea, 38-39, 176
   tea from, 162

INDEX

Lambsquarter, 34-37, 69, 147
    hints on cooking, 144,145, 146
    recipes using, 20, 21
Larkspur, 29
Larson Bay, 57
Laryngitis, 106, 128
Lathyrus maritimus, 59, 142
Lavender, tea from, 162
Laxative, 12, 19, 27, 28, 38, 119, 130
Ledum palustre decumbens, 38-39
Lemon Roll-ups, Zesty, 156
Lemonade, Rose Hip, 124
Lettuce, 76-77
Lettuce substitutes, 10, 12, 77
Lice, 118, 128
Lichen dyes, 166-167
Licorice fern, 83
Ligusticum scoticum, 52-54, 140
Lily, chocolate or Kamchatka, 41
Lily, corn, 138-139
Lily Family, 41-44, 138-139
Lingonberry, 97-98
    Chiffon Pie, 99
    dye, 171
Liver problems, 29
Loaf, Nettle, 51
Lobe, definition of, 182
Lóstochki, 71-72, 179
Lovage, beach or Scotch, 52-54
Lowbush cranberry, 97, 98
Lung troubles, 29, 81
Lupine, 59, 175
Lupine, Nootka, 141
Lupinus nootkatensis, 59, 141
Lycopodium selago, 14-15

Madder Family, 45
Magic, black 15
Mahoonah, 177
Maidenhair fern, 84
Makrétzi, 62-63
Makrétzi Soup, 63
Malína, 112, 115-117
Map, vi
Marinade for game, 39
Marinated Fireweed Shoots, 32
Marmalades, 121, 122
Maróshka, 113
Marsh fivefinger, 74
    tea from, 162
Marsh marigold, 29-30
    hints on cooking, 143, 145, 146
Matricaria matricarioides, 26-27
Mayo Clinic research, 60

Mealberry, 103
Meat dishes, seasoning for, 38, 127
Medical uses of plants, 6-7
Menstrual problems, 28, 40, 46, 75, 122
Mentha spicata, 46-47
Milfoil, 28-29
Miner's lettuce, Alaska, 71-72
Minerals in plants, 41, 48, 62, 66, 68
Mint, 61, 65
    tea from, 47, 162
    jelly from, 47, 162
Mint Family, 46-47
Mint Sauce, 47
Mogúlnik, 38-39, 176
Monastery, 3-4
Moneses uniflora, 81
Monkshood, 29, 135-136
Mordant,
    alum, 165
    chrome, 166
    copperas crystals, 166
    definition of, 182
Mordanting, 164
Mosquito repellent, 29
Mossberry, 113
Mossberry Jam, 114
Mothers, new, 27, 33, 40, 112
Mountain ash, 120-121
Mountain sorrel, 9-10, 64
    hints on cooking, 144, 145, 146
Mountain Sorrel and Fish Soup, 10
Mouth sores, 40, 56, 77, 112, 130
Muffins, 93, 102
Mulligan, Elk, 150
Muscles, sore, 45, 73
Mussels, Danny's Spicy Steamed, 155
My darlings, 136, 173
Myrica gale, 126-127
Myrica gale L., dye from, 171
Myrtle, bog, 126-127

Nagoonberry, 110-111, 179
Nagoonberry Jelly, 111
Names of plants, 1
Narcissus-flowered anemone, 136, 173
National Cancer Institute research, 128
Native
    dwellings, 48
    foods, 116, 129
    remedies, 90, 97, 104, 118, 119, 120, 126-127, 130

Nausea, 27, 46
Nerves, soothing, 27
Nettle, 48-51, 179
    dye from, 168
    hints on cooking, 145, 146
    recipes using, 14, 35, 37
    sting of, 11
Nettle
    Casserole, 51
    Loaf, 51
    Pie, Deep Dish Salmon-Wild Rice-, 161
    with Sunflower Seeds, Salmon and, 159
Niacin, 11, 34, 104
Nichols Garden Nursery, 162
Nightmares, 27
Nitrogen-fixing bacteria, 60, 129, 141
Nizanaýnik, 118-120, 180
Node, definition of, 182
Noodles from Mars, Green, 21
Nootka lupine, 141
Nosebleeds, 15
Nuphar polysepalum, 79-80
Nursing mothers, 40, 112
Nutritional value of plants, 6, 7

Octopus Salad, 161
Odoovánchik, 18-25
Omelette, Banjo Beach, 20
Onion, wild, 42-43, 138
Ooeduck Soup, 153
Oogóortsi, 43-44
Opposite, definition of, 182
Orange peel, tea from 162
Ornamental plants, 121
Ouzinkie Botanical Society, v
Ouzinkie, vi
Oxalic acid, 9-10, 11
Oxycoccus microcarpus, 98
Oxyria Digyna, 9-10

Pain-killer, 92
Paleozoic plants, 39
Pancakes, 19, 62, 93
Paparótnik, 82
Paper, 49, 131
Parasites, 11
Parasorbic acid, 120
Parmelia dye, 167
Parsley Family, 52-58, 139-140
Parsley substitutes, 20, 53
Parsnip a la Hercules, 57
Pea Family, 59-61, 141-142

Peas, beach, 59
Pectin, 87, 90
Peppermint, 46, 47
Perennial, 5, 182
Perfume, 122
Perok, 68
Perok, Salmon, 159
Petrúshki, 52-54
    Deviled Eggs with, 54
    Fish Chowder with, 157
    hints on cooking, 144, 145
    recipes using, 14, 36, 148, 151, 152, 153, 155
    tea from, 162
Phosphorous, 11, 19
Photosynthesis, definition of, 182
Picea sitchensis, 129-131
    dye from, 170
Pickling, 57
Pie, 90, 93
    Blueberry, 94
    Crowberry, 91
    Dandelion Blossom, 24
    Deep Dish Salmon-Wild Rice-Nettle, 161
    Lingonberry Chiffon, 99
    Pigweed, 35
    Salmonberry, 116
Pigweed, 34-37
Pilaf, Aleut, 148
Pills, 15
Pimples, 127
Pine Family, 129-131
Pineapple weed, 26-27, 179
    tea from, 162
Pink Family, 62-65
Pinkheads, 98
Piroshkees, Tasty Foragers', 36
Plant classification, 1-2
Plant collecting, 5
Plantago macrocarpa, 68-69
Plantago major, 66-67
    dye from, 167
Plantago maritima, 68-69
Plantain, seashore, 68-69
Plantain, broad-leaved, 66-67
Plantain, seashore, 68-69
Plantain Family, 66-69
Pleasant Harbor, 3-4
Poached Halibut, 156
Poison water hemlock, 139-140
Poison, 30, 40, 43, 55, 59, 82, 104, 119
Poisonous plants, 14-15, 134-142
Polemonium acutiflorum, 70

Polemonium Family, 70
Polemonium pulcherrimum, 70
Poléznaya travá, 4, 28, 29
Polín, 16-18, 180
Polínya, 16-18
Pollinate, definition of, 182
Polygonum alaskanum, 11
Polypodium vulgare, 83
Pond lily, 79-80, 173
Pond lily, yellow, 146
Póochki, 54-58
    dye from, 169
    hints on cooking, 145, 146
Popcorn, 80
Populus balsamifera, 131
Poque, 178
Pot Roast, Sea Lion, 147
Potassium, 11, 19, 109
Potato salad, 47
Potato substitutes,
    plants for, 75, 146
    when to gather, 144
Potatoes with goosetongue, 68
Potentilla anserina, 74-76
    dye from, 169
Potentilla palustris, 74
Potherbs, 9, 143
Potpourris, 122
Poultice, 119
    definition of, 182
Pregnancy, 18, 112
Protein, 34, 48, 56, 104
Pteridium aquilinum, 82
Pudding, Blueberry, 95
Puddings, 78, 93
Punch, Salmonberry-Highbush, 107
Purée, 144, 146
Purslane Family, 71-72
Purslane, sea, 63-65
Pussy willow, 132
Puyuk, 16, 17
Puyurniq, 110-111

Qangananguaq, 28-29

Rabies, 70
Rain flower, 71, 178
Ranunculus, 135
Rashes, 45, 81
Raspberry, 42, 61, 112-114
    American red, 112
    trailing, 113
    wild, 110-111
Red-berried elder, 104-105
Regenerate, definition of, 183

Relish, Crimson Cranberry, 102
Respiratory, definition of, 183
Respiratory problems, 66
Rheumatism, 19, 49, 119
Rhizome, definition of, 183
Rhubarb, wild, 10-14
Riboflavin, 11, 34
Rice a la Russ, Indian, 42
Rice in Gravy, Gumboot, 154
Rice substitutes, 41, 173
Riceroot, 41
Ringworm, 11
Roast, Deer, 149
Romáshka, 26-27, 179
Roots, Candied Silverweed, 75
Roots, edible, plants with, 146
Rope, 49
Rosa nutkana, 1, 121-126
Rose, wild or prickly, 121-126, 175
Rose Family, 72-76, 108-117, 120-126
Rose Hip
    Jam, 123
    Lemonade, 124
    Syrup, 123
Rose hips, 122, 176
Rose Petal Jelly, 124
Rose petals, 144, 162
Rose-Chicken Waldorf Salad, 125
Rose-topped Cheesecake, 125
Roseroot, 77-78, 144, 146, 174
Rosette, definition of, 183
Róza, 121-126
Rubus arcticus, 110-111
Rubus chamaemorus, 113
Rubus idaeus, 112
Rubus pedatus, 113
Rubus spectabilis, 115-117
Rumex, 10-14, 169
Rumex acetosella, 11, 169
Russian plant names, 2-4

Sachet, 127
Sage, tea from, 162
Salad, 78, 144
    Clover-Bright, 61
    Cockle Clam, 154
    Early Spring Mixed Wild, 69
    Franny's Favorite Spruce Island, 65
    Octopus, 161
    Rose-Chicken Waldorf, 125
    Shrimp, 151
    Spring Beauty, 72
    Summer, 20
    Wild Cucumber, 44

Salad greens, 76-77
Salads, *throughout text*
Salicin, 132
Salix, 132-133
Salmon
  and Nettles with Sunflower Seeds, 159
  Bake, Easy, 161
  Loaf, Danny's, 159
  Perok, 159
  pie, 68
  Ring with Cheese Sauce, 160
  stuffed, 59
  -Wild Rice-Nettle Pie, Deep Dish, 161
  with Garlic, Canned, 160
Salmonberries, 61, 179
Salmonberry, 115-117
  blossoms, 65, 144
  bushes, leaves under, 81
  -Highbush Punch, 107
  Jam, 117
  Pie, 116
Salts, Glauber's, 165
Salve, 45
Sambucus racemosa, 104-105
  dye from, 170
Sandwich,
  Open, 47
  Seaside, 152
Sandwich spreads, 47
Saránal, 41
Sarsaparella, 174, 176, 179
Sauce, 93
  Blender Mint, 47
  Simple Mint, 47
  Wild Fruit or Berry, 114
Saxifrage, 76-77
  hints on cooking, 144, 145, 146
  recipes using, 65
Saxifrage punctata, 76-77
Scabies, 127
Scalds, 11
Scallop Broth, 154
Scallop Recipe, Nell's Favorite, 154
Scotch lovage, 52-54
Scouring rush, 39-41
Scrubber for plates and pans, 40
Scurvy, 64, 120
  definition of, 183
Scurvy grass, 77-78
Sea Lion, Boiled, 148
Sea Lion Pot Roast, 147
Sea purslane, 63-65
Sea Soup, 158

Sea-chickweed, 63-65
Seabeach sandwort, 63-65, 144, 151
Seabeach senecio, 178
Seafood Recipes, 151-161
Seal, Cooking Instructions for, 147
Seashore plantain, 68-69
Seasoning for meat, 127
Sedum rosea, 77-78
Seed Dressing, Summer, 36
Seeds, edible, plants with, 146
Shield Fern Family, 85-86
Shíksha, 90-91
Shikshónik, 90-91
Shipóynik, 1, 121-126
Shisaki, 129-131
Shíski, 129
Shrimp
  Salad, 151
  de Jonghe, 151
Shrubs, 118-127
  dyes from, 170
Shy maiden, 81, 174
Siberian spring beauty, 71-72
Silica, 40
  definition of, 183
Silverleaf, 16-18
Silverweed, 74-76
  Cakes, 76
  dye from, 169
  hints on cooking, 146
  Roots, Candied, 75
Single delight, 81
Sinuses, stuffed-up, 28
Skin blemishes, 75
Skin diseases, 15, 60
Skin irritations, 29, 55-56, 62
Skin problems, 12, 40, 45, 128
Skin salve, 45
Skin wash, 27
Skrípka, 77-78, 174
Skunk cabbage, 139
Sleeplessness, 46
Smoking fish, 129, 131
Smoking mixture, 103
Snakeweed, 66-67
Snuff, 120
Sodium, 19, 109
Solidago lepida, 25-26
  dye from, 167-168
Solidago multiradiata, 25-26
  dye from, 167-168
Sorbus sitchensis, 120-121
Sore throat, *throughout text*
Sores, 26, 29, 40, 56, 66, 81, 112, 119, 121, 131

Sorrel, mountain, 9, 144, 145, 146
Soufflé, King Crab, 153
Soufflé, Wild Green, 37
Soup, 77
  Basic Cream, 144
  Duck, 150
  Fish Head, 158
  Makretzi, 63
  Ooeduck (Baidarka or Gumboot), 153
  Sea, 158
Soups
  plants for, 145
  plants for creamed, 146
Sourdock, 10-14, 48, 64, 69
  dye from, 169
  hints on cooking, 144, 145, 146
  recipes using, 21, 63, 65, 69
Sourdock root, tea from, 162
Sourgrass, 9-10
Spaghetti a la King Crab, 152
Spareribs, Baked Deer, 149
Spatterdock, 79-80
Spearmint, 46-47, 53
  hints on cooking, 144, 145
  tea from, 162
Spearmint Candy, 47
Species, definition of, 183
Sphagnum moss, 98
Spike moss, 14-15
Spinach, wild, 10-14
Spores, 15
Sprains, 73
Spreading wood fern, 85-86
  dye from, 169
  fiddleheads of, 144, 145
Spring beauty, 65, 178
Spring beauty, Alaska or Siberian, 71-72, 144, 145
Spring Beauty Salad, 72
Spruce, 129-131, 175
  dye from, 170
Spruce Island, vi
Spruce Island Salad, Franny's Favorite, 65
Staph infections, 119
Star of Bethlehem, 81, 116, 174
Starch source, plants for, 146
Steaks, Deer Swiss, 148
Steambath switches, 26, 127, 129
Steamed, plants to be, 145
Steep, definition of, 183
Stellaria, 62-63
Stews, 41, 42, 43, 47, 53, 150
  plants for, 145

INDEX

Stimulant, 19, 26, 119
Stinging nettle, 48-51
Stinkweed, 16-18
Stocks, plants for, 145
Stomach cramps, 29
Stomach problems, 33, 38, 66, 81, 106, 115, 119, 122, 130
Stomach ulcers, 40
Stomachaches, 46, 132
Stonecrop Family, 77-78
Storing plants, 43
Strawberries, in jelly, 105
Strawberries, wild, 179
Strawberry, beach, 108-110
Strawberry-Pineapple Conserve, Wild, 110
Streptopus amplexifolius, 43-44
String, willow bark, 132
Succulent, definition of, 183
Sugar, 56
Sulfate, iron or ferrous, 165
Sulfate mordants, 166
Sulphur, 109
Summer Salad, 20
Summer Seed Dressing, 36
Sunburn, 75
Susitna people, 104
Swamp cranberry, 98
Sweating, plants that cause, 70, 104
Sweet Cow Parsnips, 57
Sweet gale, 126-127
    dye from, 171
    hints on cooking, 145
Sweet pea, wild, 59, 142
Sweet potato, wild, 74-76
Swelling, 56
Swiss Steaks, Deer, 148
Symbiotic, definition of, 183
Syrup, Rose Hip, 123

Tabooley, 53
Talcum powder, 122
Tannic acid, 33
Tannin, 60
Tanning hides, 121
Tapeworms, 106
Taraxacum, 18-25
Tea, Mint, 47
Tea, plants for, *throughout text*
Tea recipes, 162-163
Teeth, loose, 75
Textile, definition of, 183
Thiamine, 11, 34, 104
    absorption problems, 82
Thrushes, 85

Tlingit foods, 116
Tlingit remedies, 120
Tobacco substitute, 103
Tonic, 19, 28, 49, 60, 73
    definition of, 183
    spring, 75
Tonsillitis, 120
Toothache, 29, 49, 56, 66, 75, 116
Toothpaste, 47
Toxic, definition of, 183
Trees, 128-133
    dyes from, 170
Trifolium, 60-62
Trout fillets, recipe for, 161
Tuberculosis, *throughout text*
Twisted stalk, 43-44

Ugyuun, 54-58
Ulcers, 18, 33, 40, 121, 130
Umbelliferae, 140
Upper Inlet people, 104
Urinary tract problems, 40
Urination, painful, 45
Urtica lyallii, 48-51
    dye from, 168
Usnea, dye from, 167
Uterine problems, 15
Uugaayanaq, 48-51

Vaccinium, dye from, 170
Vaccinium ovalifolium, 94
Vaccinium uliginosum, 93
Vaccinium uliginosum microphyllum, 94
Vaccinium vitis-idaea, 97, 171
Varicose veins, 17
Vascular, definition of, 183
Venereal disease, 119
Veratrum viride, 138-139
Vérba, 132-133
Veyníki, 128-129
Viburnum edule, 106
Violet, wild, 78-79, 144, 176
    blossoms of, 65, 72
    leaves of, 145
Violet Family, 78-79
Violet Jelly, 79
Vitamin, definition of, 183
Vitamin A, *throughout text*
Vitamin B, 19
Vitamin C, *throughout text*
Vitamin E, 122, 123
Vitamin content of plants, 7, 143
Vitamin deficiencies, 34
Vomiting, to cause, 12, 122

Wainiik, 128-129
Waldorf Salad, Rose-Chicken, 125
Water Lily Family, 79-80
Water for dyeing, 165
Water hemlock, poison, 139-140
Water retention, 15, 40
Watercress, 62
Watermelon berry, 43-44
Wax flower, 81, 174
Wax Myrtle Family, 126-127
Weather omens, 137
Whooping cough, 75
Whorl, definition of, 183
Wild Green Soufflé, 37
Wildwood Fritters, 69
Willow, 132-133
    inner bark of, 146
    leaves of, 144
Willow Family, 131-133
Wilted with Vinegar and Bacon, plants to be, 145
Wine, 19, 111, 116, 121, 123
    Dandelion, 24
    Elderberry, 105
Wineberry, 110-111
Winter foods, *throughout text*
Wintergreen Family, 81
Winterweed, 62-63
Wood for carpentry, 131
Worms, 66, 83
Wormwood, 16-18
Wormy berries, 93-94
Wounds, 29, 40, 81, 115, 132

Yarn for dyeing, 165
Yarrow, 4, 28-29
    tea from, 162
Yellow pond lily, 79-80, 146

Zemlyaníka, 108-110
Zesty Lemon Roll-Ups, 156
Zhólti golóvnik, 4, 25, 178
Zygadenus elegans, 138